测绘科技专著出版基金资助

三维空间关系的描述及其定性推理

Description and Qualitative Reasoning of Spatial Relation in a Three-dimensional Space

刘 新 刘文宝 李成名 著

测绘出版社

·北京·

ⓒ 刘　新　刘文宝　李成名　2010

所有权利(含信息网络传播权)保留,未经许可,不得以任何方式使用。

内容简介

　　空间关系主要包括拓扑关系、方向关系、距离关系三类,是 GIS 学科中的重要理论问题之一。它们的研究内容又可细分为空间关系描述和空间关系推理。本书阐述了三维空间关系理论方面的研究成果。首先建立了三维拓扑关系、方向关系、定性距离、综合拓扑方向关系(拓扑关系+方向关系)和位置关系的描述与表达模型。在此基础上,研究了它们的定性推理。

　　本书可作为地理信息系统专业及其相关专业研究生的选修教材或教学参考书,同时可供地理学、计算机科学、测绘科学与技术等领域从事空间信息处理与分析的研究人员和工程技术人员阅读参考。

图书在版编目(CIP)数据

　　三维空间关系的描述及其定性推理 / 刘新,刘文宝,李成名著. —北京：测绘出版社,2010.7
　　ISBN 978-7-5030-2023-0

　　Ⅰ．①三… Ⅱ．①刘… ②刘… ③李… Ⅲ．①地理信息系统－研究 Ⅳ．①P208

　　中国版本图书馆 CIP 数据核字(2010)第 099275 号

责任编辑	贾晓林	封面设计	李　伟	责任校对	董玉珍　李　艳
出版发行	测绘出版社				
社　　址	北京市西城区三里河路 50 号	电　　话		010－68531160(营销)	
邮政编码	100045			010－68531609(门市)	
电子信箱	smp@sinomaps.com	网　　址		www.sinomaps.com	
印　　刷	北京金吉士印刷有限责任公司	经　　销		新华书店	
成品规格	169mm×239mm				
印　　张	11.75	字　　数		220 千字	
版　　次	2010 年 7 月第 1 版	印　　次		2010 年 7 月第 1 次印刷	
印　　数	0001－1500	定　　价		32.00 元	
书　　号	ISBN 978-7-5030-2023-0/P・464				

本书如有印装质量问题,请与我社联系调换。

前　言

空间关系的描述和推理是当前测绘学、地理学、计算机学和认知科学等学科中共同关心的科学问题。空间关系主要包括拓扑关系、方向关系和距离关系三类，是 GIS 学科中的重要理论问题之一。它们的研究内容又可细分为空间关系描述和空间关系推理。以往空间关系的描述主要集中在二维空间关系：拓扑关系、方向关系、定性距离和位置关系的定性描述、表达及其定性推理。由于 GIS 研究范围的不断扩大，许多实际问题用二维空间关系理论无法很好地解决，研究三维空间关系理论势在必行。

本书首先建立了空间关系的描述与表达模型，在此基础上研究了空间关系的定性推理。第 1 章综合阐述和分析目前空间关系的研究现状。第 2 章采用单纯形数据模型基于点集拓扑理论研究拓扑关系的描述和推理。第 3 章研究方向关系的描述、表达模型及其定性推理。第 4 章研究距离关系的定性描述、表达及其推理。为揭示空间关系间的关联本质，第 5 章对拓扑关系和方向关系间的混合空间关系定性推理进行研究。为提高空间关系的描述、表达能力以及空间关系推理的准确性，第 6 章研究综合拓扑方向关系（拓扑关系＋方向关系）的描述、表达模型及其定性推理，第 7 章研究位置关系（方向关系＋定性距离）的描述、表达模型及其定性推理。

本书的研究工作得到了国家自然科学基金项目（40471109）、国家科技支撑计划项目子课题（2006BAJ15B02）、中国测绘科学研究院基础科研项目（77748）、地理空间信息工程国家测绘局重点实验室开放基金项目（B2808）、国家高技术研究发展计划项目（2009AA12Z147）等课题的资助。

本书的顺利完成得到了中国测绘科学研究院以及山东科技大学领导、老师和同事们的大力支持和帮助，在此一并致谢！

本书内容系作者近年来部分研究成果的总结，还有待于进一步深入研究。由于作者水平所限，书中难免有不足和谬误之处，恳请同行专家及广大读者赐教。

目 录

第1章 绪 论 ··· 1
    §1.1 国内外研究状况 ·· 1
    §1.2 空间关系的发展趋势 ·· 14
    §1.3 本书的组织与安排 ··· 16

第2章 拓扑关系描述及其定性推理 ··· 18
    §2.1 拓扑关系描述 ·· 18
    §2.2 拓扑关系定性推理 ··· 24

第3章 方向关系描述及其定性推理 ··· 57
    §3.1 二维方向关系描述及其定性推理 ·· 57
    §3.2 三维方向关系描述及其定性推理 ·· 59

第4章 定性距离描述及其推理 ··· 33
    §4.1 定性距离描述 ·· 33
    §4.2 定性距离间的定性推理 ·· 36

第5章 混合空间关系定性推理 ··· 95
    §5.1 二维混合空间关系定性推理 ·· 95
    §5.2 三维混合空间关系定性推理 ·· 102

第6章 综合拓扑方向关系描述及其推理 ··· 112
    §6.1 二维综合拓扑方向关系描述及其推理 ······································ 112
    §6.2 三维综合拓扑方向关系描述及其推理 ······································ 120

第7章 位置信息的定性描述及其定性推理 ·· 147
    §7.1 二维位置关系的定性描述及其定性推理 ··································· 148
    §7.2 三维位置关系的定性描述及其定性推理 ··································· 158

参考文献 ·· 168

CONTENTS

Chapter 1　Introduction ·· 1
　§ 1.1　Literature Review of Spatial Relations ···································· 1
　§ 1.2　Future Research on Spatial Relations ···································· 14
　§ 1.3　Framework ·· 16

Chapter 2　Description and Qualitative Reasoning of Topological Relations
　　　　　·· 18
　§ 2.1　Description of Topological Relations ···································· 18
　§ 2.2　Qualitative Reasoning of Topological Relations ··············· 24

Chapter 3　Description and Qualitative Reasoning of Directional Relations ··· 57
　§ 3.1　Description and Qualitative Reasoning of Directional Relations in 2D Space ··· 57
　§ 3.2　Description and Qualitative Reasoning of Directional Relations in 3D Space ··· 59

Chapter 4　Description and Reasoning of Qualitative Distances ················ 83
　§ 4.1　Description of Qualitative Distances ···································· 83
　§ 4.2　Qualitative Reasoning of Qualitative Distances ··············· 86

Chapter 5　Qualitative Reasoning of Mixed Spatial Relations ···················· 95
　§ 5.1　Qualitative Reasoning of Mixed Spatial Relations in 2D Space ··· 95
　§ 5.2　Qualitative Reasoning of Mixed Spatial Relations in 3D Space
　　　　　·· 102

Chapter 6　Description and Reasoning of Integrated Topological and Directional Relations ·· 112
　§ 6.1　Description and Reasoning of Integrated Topological and Directional Relations in 2D Space ··································· 112
　§ 6.2　Description and Reasoning of Integrated Topological and Directional

                    Relations in 3D Space ·· 120
**Chapter 7 Qualitative Description and Qualitative Reasoning of Location
            Information** ··· 147
   § 7.1 Qualitative Description and Qualitative Reasoning of Location
          Relations in 2D Space ·· 148
   § 7.2 Qualitative Description and Qualitative Reasoning of Location
          Relations in 3D Space ·· 158

References ··· 168

第1章 绪 论

地理信息是与研究对象的空间地理分布有关的信息,它属于空间信息,并具有区域性、多维结构和动态变化等特征。地理信息系统(GIS)是以采集、储存、管理、分析和描述整个或者部分与空间和地理分布有关的数据的空间信息系统。空间关系是 GIS 学科中的重要理论问题之一,在 GIS 空间数据建模、空间查询、空间分析、空间数据挖掘、空间推理、制图综合、地图理解及应用等过程中起着重要的作用。本章将重点分析国内外空间关系的研究状况及其存在的问题。

§1.1 国内外研究状况

对于地理空间信息,空间实体除了具有各自的几何特征和非几何特征之外,还具有空间特征实体间的相对关系,即空间关系。空间关系是空间数据组织、查询、分析和推理的基础(Frank,1992;Egenhofer,1994a;郭薇 等,1997b;曹菡 等,2001a;Chang,2002)。空间关系主要包括拓扑关系、方向关系、距离关系三类。按照研究范围,空间关系可分为二维空间关系和三维空间关系。而根据研究内容,空间关系又可分为空间关系描述和空间关系推理。

1.1.1 二维空间关系研究状况

在二维空间中,空间数据的采集、处理、表示与分析基于一个平面,即通过平面上的二维平面坐标(x,y)对研究对象进行定义,描述的是二维(two dimensions,2D)空间中的对象。空间数据可以用表达式$v=f(x,y)$表示,其中,v是点(x,y)对应的属性值(李德仁 等,1993;郭达志,2002)。在 GIS 中,经常涉及对多个空间目标的操作,使空间关系成为一项重要的研究内容。而空间关系研究的主体是空间对象之间的各种关系。空间目标大到一个国家,小到一个村镇或一个简单的地物。因此,空间目标在类型和层次上就具有多样性,这就决定了空间关系的复杂性(赵红超,2006)。

空间对象在不同的尺度下具有不同的表现形式。在大比例尺下表现为狭长面域的空间对象,在小比例尺下则表现为线状空间对象(梁启章,1995)。而面-面空间关系和线-线空间关系是不相同的。在研究空间关系时,不仅要研究原子对象之间的空间关系,还要研究原子对象与组合对象,以及组合对象与组合对象之间的空间关系,因此空间关系具有层次性。人们对空间关系的认知程度、空间数据的不确定性以及空间关系在分析过程中的不确定性等造成了空间关系具有不确定性。空间对象的空间位置和范围随着时间的变化也会发生改变,如地块的合并与分割,空间对象的时空变化会导致其空间关系也发生变化(Chen et al,2000)。空间关系的描述和推理具有完备性、严密性、唯一性和通用性(Abdelmoty et al,1994)。空间关系的研究方法主要有:决策树法、遗传算法和逻辑推理等(Mckay et al,1997)。

1. 空间实体模型

在二维空间中,点、线和面是 GIS 中最基本的实体。根据物体的复杂程度可以将物体分为简单物体(简单点、线、面)和复杂物体(复杂点、线、面)。其中,简单点是二维实数域($\mathbf{R}^2$)上仅包含一个点的连通 0D(zero dimension,0D)闭子集。简单线是在平面 $\mathbf{R}^2$ 上的有两个终点且不自交的 1D(one dimension,1D)连通闭子集,可以定义简单线为连续函数 $f:[0,1] \to \mathbf{R}^2$ 的像,并且满足 $\forall t_i, t_j \in [0,1]$,$f(t_i) \neq f(t_j), t_i \neq t_j$。简单面是平面 $\mathbf{R}^2$ 上的齐次单连通闭子集(Egenhofer et al,1991a,1991b;Clementini et al,1995a,1996;Clementini et al,1993)。

平面 $\mathbf{R}^2$ 上的复杂几何物体是简单物体的推广(Clementini et al,1995a,1996),复杂点为平面 $\mathbf{R}^2$ 上包含有限个不同点的 0D 点集。复杂线为连续函数 $f_i: [0,1] \to \mathbf{R}^2$(其中 $i=1,2,\cdots,n$)像的并,即 $f_1([0,1]) \cup f_2([0,1]) \cup \cdots \cup f_n([0,1])$。复杂面是平面 $\mathbf{R}^2$ 上的 2D 闭子集,它由多个交集为空集或在边界上相交的简单面的并构成。根据点集拓扑理论,2D 物体 A 将整个二维平面分为 3 个部分:内部(A°)、边界(∂A)和外部(A^-)(熊金城,1981)。

现实世界中存在一种有内外两个边界的物体。内边界与外边界之间的部分是一个面,称由内外边界所形成的闭集为宽边界,称具有宽边界的物体为宽边界物体。宽边界简单面是边界为宽边界的简单面,宽边界线是内部为空集的宽边界面,宽边界复杂面是边界为宽边界的复杂面(Clementini et al,1997a)。宽边界可以理解为物体的边界在一定时间内的变化范围。对一些连续变化的地理现象,内边界为最初状态的边界,外边界为最终状态的边界,如暴风雨前的海岸线和暴风雨之后的海岸线。宽边界也可以理解为物体在一段时间内的最大范围和最小范围。最大范围的边界和最小范围的边界所构成的闭区域称为宽边界,因此又称宽边界为不确定性边界。可以将宽边界物体和具有不确定性边界的物体统称为模糊物体。将不确定边界的面以及宽边界面常统称为模糊面(Dubois et al,1987;Zhan,1998)。用模糊语言描述这种不确定性是一种自然方式(Goodchild et al,1990;Couclelis,1996)。根据模糊集理论和点集拓扑理论,可用隶属函数刻画模糊面(张文修,1984)。模糊面将其所在的二维空间划分为 3 个部分:核、不确定边界和外部。核是平面 $\mathbf{R}^2$ 上隶属度值为 1 的部分,不确定边界是平面上隶属度为 $(0,1)$ 的部分,外部为平面上隶属度值为 0 的部分。当模糊物体为模糊点时,用其分布的均值和方差来描述模糊点(邓敏 等,1999)。当模糊点是运动变化的点时,分点的运动轨迹分已知和未知两种情况研究模糊点的描述(邓敏 等,2002)。宽边界有不同的近似表示方法,例如缓冲区方法、最小边界方法、凸包方法、格网方法、模糊集方法和粗糙集方法等(Brinkhoff et al,1993;Cohn et al,1995)。

2. 空间关系描述

当研究两个以上空间物体时,空间关系就必然成为研究的重要内容。空间物体的复杂多样性决定了空间关系的复杂多样性,包括由空间现象的几何特性引起的空间关系、完全由空间现象的非几何特性所导出的空间关系、由空间现象的几何特性和非几何特性共同引起的空间关系(郭仁忠,1997)。本书研究的空间关系是指由空间物体的几何特性所决定的关系。这类关系可概括为距离关系、方向关系和拓扑关系。

距离反映了物体间的几何接近程度。根据物体的空间图形,二维空间物体可抽象为点、线、面三类。根据各类物体间的组合形式,距离形式可归纳为六种:点-点、点-线、点-面、线-线、线-面、面-面间的距离。二维空间中两个区域间距离的常见定义方法有:两个区域质心间的距离,两个区域间的最短距离,两个区域间的最长距离(Santos et al,2005)。在二维笛卡儿坐标系中,两物体间的距离为两物体间的最短距离。根据人们的认知推理,距离常常映射为定性距离。定性距离必须对应于一个距离区间所指定的数值,并且能够排序和比较。用"很近、近、远、很远"等可直观地描述从近到远的距离变化(Hong,1994)。

方向关系描述一个物体相对于另一个物体的方位关系,由3个元素唯一确定:目标物、参照物和固定参考点(常指北极)。通常用指定的度数(0°,45°,…)表示原子方向,或者用具有一定方向区域的定性值或符号表示。原子方向的方向区域可用投影方法(Frank,1992,1996)或锥形系统(Peuquet et al,1987)获得。锥形系统方向区域的特点是随距离的增大而增大。由于锥形方法没有考虑参照物的大小和形状对方向划分的影响,因此当参照物为非点状物体时,采用锥形方法具有一定的局限性,故实用上通常采用投影方法划分方向区域。经常使用的方向关系有:东、南、西和北四方向关系。如果细分,可增加东北、西北、东南和西南,从而形成八方向关系。如果再细分,也可以定义更细的方向关系,如16个不同的定性方向。还有一种方向关系叫位置方向关系,如上、下、前、后、左和右等。根据Frank(1996),基于投影的模型与基于锥形的模型相比,具有以下优点:基于投影模型产生的推理结果比基于锥形模型产生的推理结果更精确,并且比锥形模型易于实现。但锥形模型易于对空间进行更细的划分(Goyal,2000)。显然,方向关系不具有旋转不变性。在方向关系的描述方法中,首先是采用锥形方法或投影方法将参照物所在的空间进行划分,得到4个方向区域和1个同一区域或8个方向区域和1个同一区域。对于后者,常采用九交矩阵模型(9IM)表示方向关系。当隶属函数为布尔函数时,9IM为粗略方向关系矩阵(Goyal,2000);而当隶属函数为几何度量值的比值时,9IM 为详细方向关系矩阵(Goyal et al,2001)。

拓扑关系是在连续空间变换(旋转、放大、缩小等)下保持不变的关系(Santos et al,2005)。拓扑关系不考虑空间物体的度量和物体在空间中的方向(郭仁忠,1997)。由于拓扑关系的改变意味着其他几何特征的改变,因此拓扑关系是一个非常重要的空间信息,吸引了众多学者来研究(Egenhofer et al,1991a,1991b)。

3. 拓扑关系描述方法

(1)点集拓扑方法。GIS 领域中广泛使用的拓扑关系描述方法是建立在点集拓扑学之上的四交模型(4IM)和九交模型(9IM)。根据点集拓扑学,空间对象所在的平面可以分为内部、边界和外部 3 个子集(熊金城,1981)。4IM 是根据平面 R^2 上两个空间对象的内部和边界的交集所组成的 2×2 矩阵来判断对象间的拓扑关系,称这个 2×2 矩阵为四交矩阵(Egenhofer,1993)。若物体为精确物体,通常用空集(0)和非空集(1)表示交集的值。因此,4IM 能够区分 $2^4=16$ 种拓扑关系,除了不可能实现的情况之外,4IM 能够区分 8 种面-面关系、11 种线-面关系、12 种线-线关系、3 种点-面关系、3 种点-线关系和 2 种点-点关系。两个模糊物体的四交矩阵是模糊矩阵(Bjorke,2004)。9IM 是用两物体的内部、边界和外部间的交所组成的 3×3 矩阵来描述两物体间的拓扑关系,所得 3×3 矩阵为九交矩阵(Egenhofer et al,1991b)。对于精确物体,通过考虑 9 个交集的值:空集(0)和非空交集(1),使九交矩阵理论上能够区分 $2^9=512$ 种关系。但实际上,对于精确边界的简单区域只有 8 种面-面拓扑关系可以实现(disjoint,meet,overlap,coveredby,inside,cover,contain,equal),有 36 种线-线拓扑关系可以实现,19 种线-面拓扑关系可以实现。这些关系是相互排斥并且是完全覆盖的(Egenhofer et al,1991a)。利用 9IM 还可以研究复杂线-线间的拓扑不变性(Clementini et al,1998)。对于宽边界的简单物体,9IM 用宽边界替换九交矩阵中的精确边界,利用九交矩阵可以区分平面 R^2 上宽边界简单面间的 44 种拓扑关系,宽边界复杂面间的 56 种拓扑关系,这其中有 44 种拓扑关系与宽边界简单面间的拓扑关系是一致的(Clementini et al,2001),还可以区分不确定线间的 146 种拓扑关系(Clementini,2005)。对于模糊物体区域间的拓扑关系,由于模糊物体边界的不确定性导致物体间的拓扑关系是模糊的。模糊物体 A 的截 $A_\partial(0 \leqslant \partial \leqslant 1)$ 是精确物体,用 A_∂ 的内部、边界和外部的九交矩阵研究模糊物体间的拓扑关系(刘文宝 等,2000)。

(2)DEM 方法。DEM 方法是在 4IM 或 9IM 中引入交集的维数来区分拓扑关系的一种方法。设定空集的维数为 -1,点的维数为 0,线的维数为 1,面的维数为 2(Clementini et al,1993)。把维数引入 4IM(即 DE+4IM),理论上有 256 种不同情况。实际上除了不可能关系之外,存在 12 种面-面拓扑关系、17 种线-面拓扑关系、3 种点-面拓扑关系、24 种线-线拓扑关系、3 种点-线拓扑关系和 2 种点-点拓扑关系。把维数引入 9IM(即 DE+9IM)不仅可以区分 0D-meet、1D-meet、0D-cover、

1D-cover、0D-overlap、1D-overlap 等拓扑关系,还可以区分 12 种面-面拓扑关系、31 种线-面拓扑关系、3 种点-面拓扑关系、36 种线-线拓扑关系、3 种点-线拓扑关系和 2 种点-点拓扑关系(Clementini et al,1995a)。

(3)最小边界矩形法。最小边界矩形(Minimum Bounding Rectangles,MBRs)是与 X,Y 轴平行的且与物体相切的矩形。由于外接矩形比空间对象简单,可以降低空间关系处理的复杂度,因而可用两空间物体最小边界矩形间的拓扑关系来近似逼近研究物体间的拓扑关系。在具有拓扑约束的查询中,用 MBRs 间的拓扑关系作为快速发现 R 树所描述的拓扑关系的标准(Papadias et al,1995),用 MBRs 间的拓扑关系优化查询方法(Clementini et al,1994)。

(4)区域连接法(Region Connection Calculus,RCC)。如图 1.1 所示,首先定义两个区域之间的二元连接关系 $C(A,B)$,把 $C(A,B)$ 的自反性和对称性作为公理,用量词 $\forall$ 和 $\exists$ 定义两个区域之间的 8 种基本拓扑关系 RCC-8(又称为 RCC-8 谓词):DC(相离),EC(外切),PO(部分相交),EQ(相等),TPP(内切),TPPi(被内切),NTPP(被包含),NTPPi(包含)(Randell et al,1992)。对二元关系 $C(A,B)$ 有两种不同的定义:一种是面 A 和 B 至少有一个共同点(Clarke,1981,1985),另一种是面 A 和 B 的拓扑闭包至少有一个公共点。(Cohn et al,1997;Bennett,2000) RCC-8 谓词仅能作用于原子区域,不能作用于原子区域的交集、并集和补集。在区域上执行布尔操作将提高 RCC-8 的表达能力,将 RCC-8 扩展为 BRCC-8(Renz,1998)。这样,BRCC-8 谓词不仅能作用在原子区域,也能作用在原子区域的交集、并集和补集(Wolter et al,2000)。

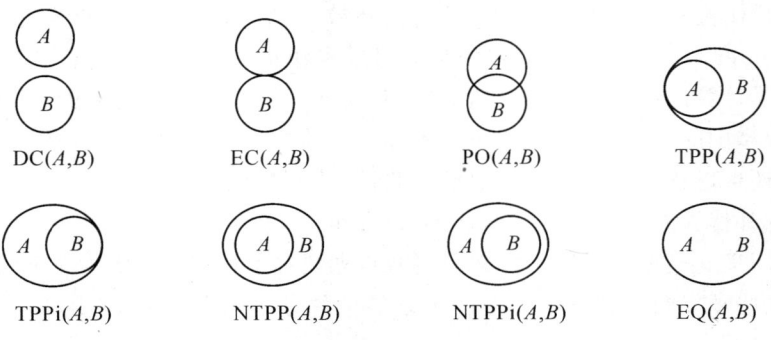

图 1.1 区域间的 8 种拓扑关系 RCC-8

(5)CBM(calculus-based method)方法。CBM 是基于计算描述拓扑关系。CBM 定义了 5 种拓扑关系 TR={touch,in,cross,overlap,disjoint}和边界算子 b(A)。当 A 为面时,b(A)为圆环;当 A 为线时,b(A)为线的端点。CBM 具有 3 个优势:完备性、排他性、表达力强。完备性是指拓扑关系能够描述所有可能的拓扑情况,即对于任意两个空间物体 A 和 B,总存在 CBM 定义的一个拓扑关系 $r\in$

TR,使得 A 和 B 间具有拓扑关系 r,即 $r(A,B)$ 成立。排他性是指两个空间物体之间仅有一种拓扑关系,即 $r_1,r_2 \in$ TR, $r_1 \neq r_2$, $r_1(A,B)$ 和 $r_2(A,B)$ 仅有一种情况成立。点、线和面之间的这 5 种拓扑关系是相互排斥并且能够覆盖所有的拓扑情况,且 CBM 能表达 DEM 的所有情况(Clementini et al,1993)。DEM 与 9IM 结合之后与 CBM 具有相同的表达力(Clementini et al,1995a)。

(6) 2D-STRING 法。2D-STRING 法又称符号投影法,是由 Chang 等提出的二维字符串模型(Chang et al,1987,1988)。2D-STRING 把对象在 X 轴和 Y 轴上的投影坐标利用"=","<",":"三个算子连接起来形成一个字符串。其中,"="表示两个物体沿某一坐标轴有相同的位置,"<"表示两物体沿坐标轴有"左-右"或"上-下"关系,":"表示两物体有相同位置。由于 2D-STRING 间关系算子所能表达的空间关系很有限,许多学者对 2D-STRING 进行了改进(Jungert,1988;Jungert et al,1989)。但 2D-STRING 仍无法描述关于一条斜线的坡度信息。Jungert(1992)扩展了 2D-STRING,将投影平行于斜线方向而不是平行于水平和垂直坐标轴形成了 2D G-STRING,孙玉国(1993)将原 2D G-STRING 进行了扩展,形成了 2D T-STRING。2D T-STRING 能够处理三类定性空间关系:拓扑关系、顺序关系、辅助关系(包围与半包围关系)。2D-STRING 不仅可以描述拓扑关系也可以描述方向关系。

(7) Voronoi 图法。Voronoi 图是一种动态数据结构,按照距每一目标最近的原则将连续空间分为若干个 Voronoi 区域,每个 Voronoi 区域只包含一个目标(Franz,1991;李成名,1998;陈军,2002;赵仁亮,2002)。Egenhofer 等研究者所提出的 9IM 把目标的补作为外部区域,导致内部、边界和外部的线性依赖关系(Chen Jun et al,2001)不利于区分空间邻近和相离关系。同时,空间目标的补太大,带来了计算的复杂性。针对这一特点,陈军 等用目标对象的 Voronoi 区域代替九交模型中的外部区域,得到 V9I 模型。V9I 模型不仅可以区分空间邻近和相离关系,还可以描述复杂对象间的拓扑关系,从而提高空间关系的分辨率,克服了以目标的补作为外部所带来的不足(Chen Jun et al,2001)。

(8) 广义交模型描述法。广义交模型是先把对象和空间分解为典型的子集,然后利用这些子集间的交来描述空间关系的一种方法(Abdelmoty et al,1995)。设 x 为一对象, x 所在的空间为 X,且 $x_1,x_2,\cdots,x_n$ 为 x 的子集,则 $x = \bigcup_{i=1}^{n} x_i$, x 的补集为 $\bar{x} = \bigcup_{i=n+1}^{m} x_i$, $X = x \bigcup \bar{x}$。利用上面的分解,对象 x 和 y 间的关系可以定义为

$$R(x,y) = (\bigcup_{i=1}^{n} x_i) \bigcap (\bigcup_{j=1}^{m} y_j)$$
$$= x_1 \bigcap y_1, x_1 \bigcap y_2, \cdots, x_1 \bigcap y_m, x_2 \bigcap y_1, \cdots, x_2 \bigcap y_m, \cdots, x_n \bigcap y_1, \cdots, x_n \bigcap y_m$$

式中, $x_i \bigcap y_j$ 可以为空集或非空集。这个交集可以用一个 $n \times m$ 矩阵来表示。通

过不同的空间分解策略,广义交模型可以描述复杂对象间的空间关系。

(9)符号阵列法。符号阵列(symbolic array)法是用符号阵列描述空间物体间的空间关系。其基本思想是用阵列元素描述空间物体,用元素间的空间关系描述物体之间的关系(Papadias et al,1992)。符号阵列法的不足之处:一种情况可能对应多个符号阵列。为了保证描述的唯一性,Papadias 和 Sellis(1994)提出了用空间索引描述拓扑和方向关系,而不考虑大小、形状和距离等信息。

4. 拓扑查询

空间关系查询主要是具有拓扑约束的空间查询。拓扑查询主要解决两个问题:一是查询与给定的空间实体具有拓扑关系的空间实体,二是查询两个给定的空间实体之间具有什么拓扑关系。对于拓扑查询可分为精确物体的空间查询和模糊物体的空间查询。精确物体的空间查询又分为精确面-面、面-线、面-点、线-线、线-点和点-点查询(Clementini et al,1994,1995b)。模糊物体的空间查询也分为上述 6 种查询(Clementini et al,2001;Beauboef et al,2004)。查询语言设计中要考虑自然查询语言中的不确定性和模糊性问题,并研究如何把模糊和不精确性语言转化为自然查询语言(Wang,2003)。

5. 拓扑关系的动态变化

拓扑关系是与时间相关的,并随自然现象和人为因素而发生改变。现有文献主要利用最近拓扑关系图研究拓扑关系的动态变化。最近拓扑关系图是把拓扑关系组织在一个图中,每个关系有一个节点,并且有一个弧将具有最小距离的每对拓扑关系相连(Egenhofer et al,1992)。拓扑关系的最小距离是对应的四交(或九交)矩阵中不同值的个数。对该图的另一个解释是把两个关系之间的弧作为从一个关系到另一个关系的平滑过渡,称之为概念领域图(Freksa,1992),也称之为连续网络(Cui et al,1993)。概念邻域图分为点-点概念领域图、点-线概念领域图、点-面概念领域图、线-线概念领域图、线-面概念邻域图、面-面概念领域图(Clementini et al,1997a, 2001;Cohn et al,1997)。

6. 空间关系推理

空间关系推理是利用测量、观察或推理所获得的物体在空间中的信息以及物体之间的关系,推导隐含在物体之间的空间关系,并得出有效结论的过程(Sharma,1996)。空间关系推理是继空间关系描述后的又一个重要研究内容。空间关系推理分为两个层次(Hong,1994):在第一层次上,根据空间关系的建模方法将空间关系推理分为直接模型化推理和间接模型化推理;在第二层次上,进一步将直接模型化和间接模型化推理分为定量推理和定性推理。定量推理是利用数值计算解决推理任务,定性推理是用逻辑或推理规则等工具得出推理结果。

间接模型化空间关系定性推理包含有:图标索引法(iconic indexing)(Chang et al,1987;Jungert, 1992)、Allen 区间逻辑及其扩展法(Allen,1983;

Guesgen,1989;Simmons,1990;Struss,1990;Vilain et al,1990;Montanari et al,1993)和符号阵列法(Papadias et al,1992,1994)。其中,基于 Allen 区间的时态逻辑在定性推理的研究中经常被采用。Allen 区间的最初提出是为了描述在一维时间域上,两个任意时间区域(段)之间的所有可能关系。两个时间段的所有可能关系共有 13 种,如图 1.2 所示(Hong,1994)。后来,Guesgen(1989)将 Allen 时态逻辑扩展到高维,进行空间推理。间接模型化空间关系定量推理有笛卡儿坐标计算等。

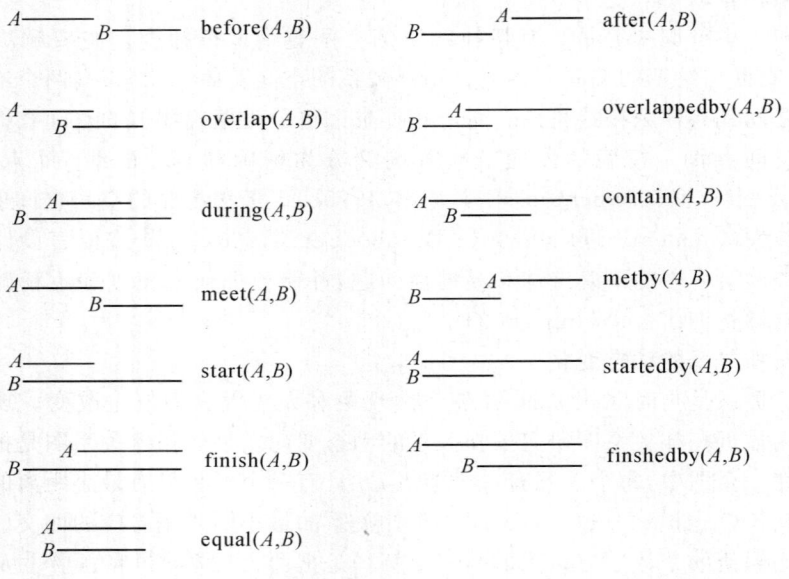

图 1.2 Allen 的两个时间段之间的 13 种可能关系

直接模型化空间关系的定量推理有:TOUR 模型推理(Kuipers,1990a,1990b;Kuipers et al,1990)、SPAM 模型推理(McDermott et al,1984)、MERCATOR 模型推理和 FUZZY 推理(Dutta,1989a,1989b)。按照所涉及的空间关系类型,直接模型化空间关系的定性推理可分为同类空间关系推理、异类空间关系推理、混合空间关系推理、综合空间关系推理(Sharma,1996)。同类空间关系推理是指推理过程中仅使用一种空间关系推理,如从拓扑关系推理拓扑关系,从方向关系推理方向关系。由于点、线、面间拓扑关系是不同的,所以同类拓扑关系推理又可进一步分为单种空间对象类型的推理,如两个线对象或两个面对象间的推理(Papadias et al,1999),以及多种空间对象类型的推理。设空间物体 A 与 B 的空间关系是 R_1,B 与 C 的空间关系是 R_2,A 与 C 的空间关系是 R_3。异类空间关系推理是由两种不同类型的空间关系 R_1 和 R_2,推导出空间关系 R_3,并且 R_3 的类型是 R_1 的类型或者是 R_2 的类型,如根据拓扑关系 R_1 和方向关系 R_2,推导 A 与 C 的拓扑关系或方向关系。混合空间关系推理是根据一种空间关系推导另一种空间关系的推理,如根据拓扑关系推导方向关系或定性距离,根据方向关系推导拓扑关系或定性距离,根据定性距离推导方向关系或拓扑关系。其中根据方向关系推导拓扑关系可以分为两种类型(杜世宏,2004):基于一个参照系的推理和基于两个参照系的推理。基于一个参照系的推理是指已知 A、B 及 A、C 间的方向关系 R_1 和 R_2,推理 B、C 间的拓扑关系。此时,R_1 和 R_2 都是以 A 为参照对象,故称为基于一个参照系的推理。基于两个参照的推理是指已知 A、B 及 B、C 间的方向关系 R_1 和 R_2,推理 A、C 间的拓扑关系。此时,R_1 和 R_2 分别以 A 和 B 为参照对象,故称为基于两个参照系的推理。

综合空间关系推理是使用两种以上的空间关系进行推理。综合空间关系推理分为综合两种空间关系推理和综合三种空间关系的推理。综合两种空间关系的推理主要是综合拓扑关系和方向关系推理以及综合定性距离和方向关系的推理,其中,综合拓扑关系和方向关系的推理(Hernár.dez,1994;Sharma,1996;Sharma et al,1995;杜世宏,2004)是根据目标对象 B 与参照物 A 的拓扑关系和方向关系,以及目标对象 C 与参照物 B 的拓扑关系和方向关系,推导目标对象 C 与参照物 A 的拓扑关系和方向关系(Hernández,1990,1994)。综合定性距离和方向关系的推理(Down et al,1973;Frank,1992;Clementini et al,1997b;Sharma,1996)是根据目标对象 B 与参照物 A 的定性距离和方向关系,以及目标对象 C 与参照物 B 的定性距离和方向关系,推导目标对象 C 与参照物 A 的定性距离和方向关系。Frank(1992)提出了一种适于距离和方向关系推理的代数,该代数包含三个部分:一是距离和方向关系符号,如距离符号"近"和方向关系符号"北";二是运算集(逆,复合等);三是公理集(推理规则)。距离方向关系的定性推理又称为位置信息的定性推理。在欧氏几何和笛卡儿坐标下,点的位置描述为点的坐标。在欧氏空间中,把连接两点的矢量视为两点的位置关系。在适当的尺度下,矢量的长度和方向对应位置关系的距离和方向。位置关系包含两个关系(Down et al,1973):距离关系和方向关系。当距离关系和方向关系均为定性关系时,位置关系就是定性的。称由 A 与 B 的定性位置关系和 B 与 C 的定性位置关系推导 A 与 C 的定性位置关系的推理为位置关系的定性推理。综合三种空间关系推理是综合拓扑、方向和定性距离三种空间关系的推理(Santos et al,2005)。拓扑、方向、距离三类空间关系之间并不是完全相互独立的,它们之间存在着一定的联系。将拓扑、方向、距离三种关系结合起来整体考虑,分析它们之间潜在的相互影响,构造一种拓扑、距离、方向关系的集成描述和推理模型,将提高空间关系描述的唯一性和空间关系推理的准确性(曹菡 等,2001b;何建华 等,2004)。由于定性推理在一定程度上比定量方法更容易被人们接受,因此定性推理成为空间推理的主要方法(郭平,2004)。上述空间推理分类用图 1.3 表示。

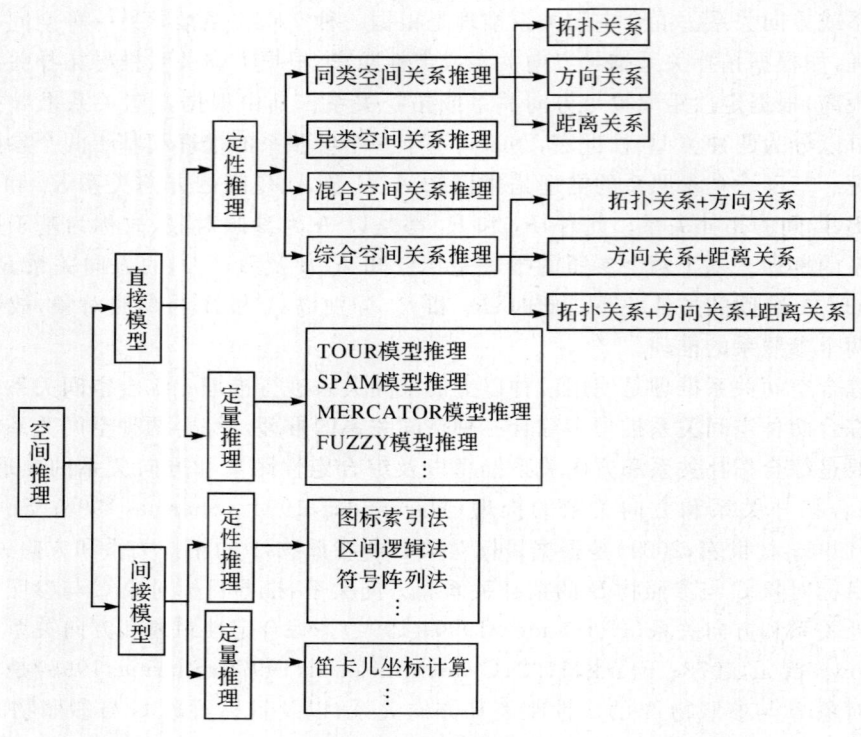

图 1.3 空间推理分类

按空间关系对是否考虑时间模型,将空间关系推理分为时空关系推理(谢琦,2006;孙海滨,2006)和非时空关系推理(Hong,1994;Sharma,1996)。按照所涉及的层次,可分为单层空间推理和多层空间推理(层次空间关系推理),多层空间推理是单层空间推理的推广,是在多个概念层上研究空间推理(Timpf,1992;Clementini et al,1994,2000)。

空间关系的推理方法可分为组合表法(Papadias et al,1999)、代数法(Frank,1992,1996)、逻辑法(如一阶逻辑法(Clarke,1981;Cui et al,1992,1993;Randell et al,1992;Bennett,1997))、时态逻辑法(孙海滨,2006)、语意网络推理(曹菡,2002)、基于产生式的推理(孙玉国,1993)、基于知识的推理等。曹菡(2002)提出代数推理、组合表推理和产生式推理的混合推理方法,能充分发挥各种推理方法的优势。

对空间关系的研究一个重要方面还表现在对空间联结规则的研究(Agrawal et al,1994;刘新 等,2007a)。联结规则是形如 $X \rightarrow Y(support, confindence)$ 的规则,其中 $support$ 和 $confindence$ 分别为 $X \rightarrow Y$ 的支持度和可信度。空间联结规则是在联结规则的前件或后件中包含空间关系的联结规则(Koperski et al,1995),在多个概念层次上挖掘联结规则叫多层联结规则挖掘(Han et al,1995),前后件中包含时空关系的联结规则为时空联结规则(Mennis et al,2005)。研究联结规则的算法主要有 Apriori 算法(Agrawal et al,1993),后来 Beauboef 和 Ladner 扩展了 Apriori 算法。当事务集为粗糙集时,采用粗糙支持度和粗糙可信度评价联结规则(Beauboef et al,2004)。事务集为模糊事务集时,采用模糊支持度和模糊可信度评价联结规则(Ladner et al,2003)。

7. 空间关系推理规则的表示

目前见到的空间关系推理规则表示方法主要有基于产生式的表示、组合表表示、代数表示和基于逻辑的表示。其中,产生式表示规则是以"IF THEN"的形式表示;组合表表示是采用二维表格存储表示所有可能的推理结果的一种规则表示形式,它实质上是一种基于框架的知识表示,用二维表的数据结构来实现,其优点是简洁、直观、查询方便,其规则是根据行 $R(A,B)$ 和列 $R(B,C)$ 演绎出有关 A 和 C 的关系事实 $T(A,C)$,如二维空间中面-面拓扑关系推理组合表(Egenhofer,1994),该表 8 行 8 列共 64 种组合,任意两对面对象拓扑关系的推理可以在组合表中按照行 $top(A,B)$、列 $top(B,C)$ 进行查找,就可得到 $top(A,C)$ 的可能结果集;代数表示是用代数运算符号表示推理规则;基于谓词逻辑的表示是利用命题演算、谓词演算等方法来描述一些常识性的知识,并根据现有事实推出新事实的方法。

1.1.2 三维空间关系研究状况

1. 三维空间的空间数据模型

三维(three dimensions,3D)空间是用三维坐标系统 (x,y,z) 定义的,(x,y,z) 表示三维空间中某点的坐标值,它表示一个空间位置。三维空间的数据可以用表达式 $v=f(x,y,z)$ 表示,其中,v 表示点 (x,y,z) 的属性值(万剑华,2001;郭达志,2002)。三维空间数据模型是关于三维空间数据组织的概念和方法,它反映了现实世界中三维空间实体及实体间的相互联系。根据模型所具有的主要特征,三维空间数据模型大致归纳为四类(万剑华,2001):矢量模型、栅格模型、混合或集成数据模型、面向对象数据模型。其中,三维矢量模型是二维点线面矢量模型在三维中的推广,它将三维空间中的实体抽象地分为三维空间中的点、线、面、体四种基本元素,然后以这四种基本几何元素的集合来构造更复杂的对象。代表性的数据模型有:边界表示法、基于 3D FDS(formal data structure)的数据模型(Moleannar,1992)、顾及空间剖分的三维拓扑数据模型(郭薇,1998;陈军 等,1998b)、面片模型、四面体格网模型、n 组拓扑关系的矢量数据结构、城市 3D GIS 矢量数据模型。其中,基于 3D FDS 的数据模型,仅适于表达规则简单的空间实体,难以表达地质及环境领域中没有规则边界的复杂三维空间实体。四面体格网模型(Franz,1991;

Raper et al,1991)仅考虑了空间实体内部结构的划分,没有考虑空间表面形态,难以表达三维面状目标及线状目标(陈军 等,1998b)。顾及空间剖分的三维拓扑数据模型兼顾了基于栅格的数据模型和基于矢量的数据模型两种表示方法的优点,该模型不但可以有效地表达规则形状的空间实体,还可以有效地表达不规则形状的空间实体(万剑华,2001)。

按照拓扑学理论,任意一个三维空间实体是一个可定向的 n 维伪流形(Armstrong,1997),它对应于一个具有良好单纯形结构的 n-单纯复形,在几何上可剖分成若干个维数小于或等于它的、连通但不相互重叠的 k-单纯形($k \leqslant n$)(郭薇 等,1997a)。0-单纯形、1-单纯形、2-单纯形、3-单纯形和平面是组成三维空间实体的五种基本几何元素。0-单纯形、1-单纯形、2-单纯形、3-单纯形所对应的几何图形分别为点、线段、三角形和四面体。平面是所有具有相同法矢量的 2-单纯形的集合,用于减少数据冗余及更有效地表达具有规则边界的面状实体和体状实体。称用点、线段、三角形、四面体这些单纯形表示空间实体的数据结构为单纯形数据结构(simplicial data structure,SDS)(Christopher,1995;易善桢,1999)。根据点集拓扑学(熊金城,1981),0-单纯形是三维空间中的 1 个点,其边界为空;1-单纯形为闭线段,其边界为闭线段的 2 个端点;2-单纯形为三角形,其边界为三角形的 3 条边;3-单纯形为四面体,其边界为四面体的 4 个面,四种单纯形如图 1.4 所示。

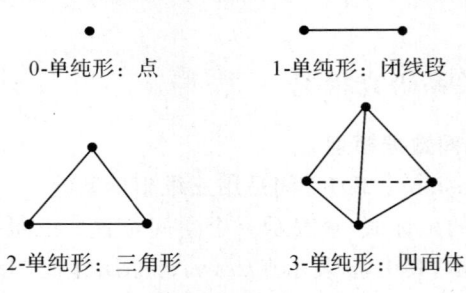

图 1.4 四种单纯形

在三维空间中用 (x,y,z) 表示点,用节点 $(x_i,y_i,z_i)(i=1,2)$ 表示线段的两个端点,用节点 $(x_i,y_i,z_i)(i=1,2,3)$ 表示三角形的三个顶点,用节点 (x_i,y_i,z_i) $(i=1,2,3,4)$ 表示四面体的四个顶点(陈军 等,1998b;易善桢,1999)。由于单纯形是凸的,一个点 P 落在 k-单纯形内当且仅当它可以写成线性组合 $P=\lambda_1 P_1+\lambda_2 P_2+\cdots+\lambda_{k+1}P_{k+1}$,其中,$P_1,P_2,\cdots,P_{k+1}$ 为 k-单纯形的顶点,$\lambda_i > 0, i=1,2,\cdots,k+1$,并且 $\lambda_1+\lambda_2+\cdots+\lambda_{k+1}=1$,所以用点描述 0-单纯形,线段的两个端点描述 1-单纯形,三角形的三个顶点描述 2-单纯形,用四面体的四个顶点描述 3-单纯形。

用 SDS 作为 3D-GIS 空间表示,原因有(易善桢,1999):①它可以用一致的模型表示点、线、面、体;②表面由三角形来构造,体由四面体来构造,三角形由最少的点和边组成,四面体由最少且固定数量的三角形、边、点组成,三角形和四面体均为凸的,把体分解为单纯形,它包含了最简单的几何特征;③在地质学中,常有不规则测量点分布,它们作为规则的栅格表达往往是不精确的,不规则测量可用三角网来插值,并且不规则三角形网表示适合地质地层模型和断层的表示。这种使用代数目标表示拓扑空间的结构,其形式更符合计算操作,有利于空间推理算法的开发(曹菡,2002)。同时,采用单纯形把几何体分解成子几何结构,在这种几何结构上执行操作,使计算变得非常容易。

2. 三维空间拓扑关系的描述方法

陈军 等(1998a)在 Egenhofer 等人提出的二维空间实体拓扑关系九元组描述框架(Egenhofer et al,1995)和 Clementini 等人提出的维数扩展法研究基础上,用 DE+9IM 方法研究三维空间中单纯形间的拓扑关系。两单纯形交集的维数为 0、1、2、3,特别当交集为空时,设交集的维数为 -1,则用 DE+9IM 可以区分 5^9 种拓扑关系,对 5^9 种拓扑关系中有意义并可以实现的拓扑关系进行分类共有 6 种拓扑关系:相邻(touch)、包含(in)、相交(cross)、部分覆盖(overlap)、相离(disjoint)和相等(equal)(陈军 等,1998a),这 6 种拓扑关系具有完备性和互斥性。根据点集拓扑理论,三维空间的任意一个实体,在几何上均可以剖分成若干个 k-单纯形($0 \leqslant k \leqslant 3$)。设有两个三维空间实体 L 和 M,分别由 p 和 q 个 k-单纯形组成,即 $L=\{L_j^k|0\leqslant k\leqslant 3,1\leqslant j\leqslant p\}$,$M=\{M_i^k|0\leqslant k\leqslant 3,1\leqslant i\leqslant q\}$,用矩阵

$$\boldsymbol{T} = \begin{bmatrix} T(L_1^k,M_1^k) & \cdots & T(L_1^k,M_q^k) \\ \vdots & & \vdots \\ T(L_p^k,M_1^k) & \cdots & T(L_p^k,M_q^k) \end{bmatrix} \qquad (1.1)$$

描述三维空间中实体 L 和 M 间的拓扑关系。

采用 OpenGIS 的数据模型,用 9IM 方法可以区分并能实现的线-线拓扑关系共有 33 种,线-面 31 种(表 1.1)、线-体 19 种(表 1.2)、面-面 38 种、面-体 19 种、体-体 8 种拓扑关系(朱铁稳,2002;Zlatanova,2009)。

表 1.1 线-面拓扑关系

图形						
序号	1	2	3	4	5	6
图形						
序号	7	8	9	10	11	12

图形						
序号	13	14	15	16	17	18
图形						
序号	19	20	21	22	23	24
图形						
序号	25	26	27	28	29	30
图形						
序号	31					

表 1.2　线-体拓扑关系

图形						
序号	1	2	3	4	5	6
图形						
序号	7	8	9	10	11	12
图形						
序号	13	14	15	16	17	18
图形						
序号	19					

§1.2　空间关系的发展趋势

近年来,随着空间对地观测技术的发展,三维地理空间信息获取与更新的能力有了飞速发展,海洋、环境等地学应用领域已逐步形成多维空间信息源(陈述彭,1997;李德仁,1997a;Wright et al,1997)。与此同时,随着空间数据库更新的技术方法和生产流程不断发展,一些城市、矿山等 GIS 部门开始更新原有的二维(x, y)数据和扩充第三维(z)数据,并逐步积累能反映地理空间要素时空分布的多维(x,y,z,t)数据源。另外,随着数字地球概念的普及和空间数据基础设施建设的深入,人们迫切需要在具有多分辨率数字表达的三维地球上集成、管理与地球有关的海量空间数据及相关信息,使其成为人们认识、改造和保护地球的重要信息源(Gore,1998)。就基础地理数据本身而言,迄今为止人们一直是按照平面图或铺盖数据模型,将具有鲜明的多维、动态特征的现实空间世界抽象为二维、静态目标。

在二维空间中,空间数据的采集、处理、表示与分析是基于一个平面,即通过平面上的二维平面坐标(x,y)来定义研究对象,描述的是二维对象(万剑华,2001)。事实上,我们生活在一个现实的三维世界中,任何空间现象都是三维的,应该用一个三维坐标系统(x,y,z)来描述。在三维空间中 x、y、z 三个坐标轴是相互独立互不相关的。一个平面坐标(x_0,y_0),对应的 z 值是一条直线 $\begin{cases} x=x_0 \\ y=y_0 \end{cases}$。用 2D GIS 描述三维空间对象本身就存在着近似性。

今后的发展趋势是从研究二维地理空间框架数据向研究多维、动态地理空间框架数据发展,以便向用户提供真三维的基础地理数据(陈军,2003)。以往,"应用驱动"和"技术导引"是国内外 GIS 发展的主要驱动力,但随着 GIS 应用的深入与产业化发展,对 GIS 基础理论的需求越来越迫切(李德仁,1997b;Moleannar,1989)。为此,国内外许多 GIS 理论研究机构纷纷成立,例如美国国家自然科学基金委员会(NSF)资助成立了美国国家地理信息与分析中心(NCGIA)等,重点研究空间关系、空间推理、空间认知、数据模型、地理信息机理和地理信息不确定性等(Goodchild,1995)。我国也成立了国家基础地理信息中心,并提出了构建"数字中国"地理空间基础框架的发展目标,以便为政府和社会提供三维空间定位数据和基础地理信息服务(陈军,2003)。近年来,我国在多维动态方面的相关研究主要集中在数据模型、动态处理、时空分析和可视化,以及空间数据的组织与查询(陈军,2000,2004)。对多维动态空间数据处理的研究主要是三维(x,y,z)数据模型、二维空间实体及其动态变化(x,y,t)的时空建模等。缺乏系统深入地对多维空间数据模型的建立、动态处理和时空分析等方面的一些关键理论和技术方法等方面的研究。当前,国家已加大了对多维动态数据的研究支持,包括支持研究多维空间数据模型与建模方法、空间数据动态处理与集成管理、多维空间数据时空分析与可视化等方面的基本理论和关键技术,以期加强对三维空间实体及其时空变化的四维数据建模(4D data modeling)的理解和认识。多维动态数据模型是数字地球基础研究方面的一个核心问题,关系到如何全面、详尽地描述人们居住的地球,以便有效地保护我们的生存环境(陈军 等,2000)。

自 20 世纪 70 年代以来,GIS 在气象、环保、灾害、海洋、生态等方面的研究不断深入,在地学等涉及全球变化或地下工程(如矿山、地质、隧道)领域的应用不断扩展。GIS 所涉及的范围已经向上进入大气层及地球的外层空间,向下进入地球内部。以矿山 GIS 为例,其空间对象包括上至大气环境,中至山川植被、建筑道路,下至矿坑边坡、井巷工程、煤岩地层等(吴立新 等,2005)。在这种情况下,许多实际问题用 2D GIS 不能很好地解决,必须对地理现象进行三维描述和分析,研究 3D GIS 已势在必行(李清泉,1998)。国内外在 3D GIS 方面的研究进展迅速,三维拓扑分析、三维空间分析等,已经成为许多学科交叉的技术前沿和攻关热点,相应的理论、技术、方法与软件系统得到不断丰富和发展(吴立新,2004)。由于空间关系理论及其研究直接影响着 GIS 系统的设计、开发和应用,因而受到国际 GIS 及相关学术界的高度重视(陈军 等,1999)。

由于空间问题固有的复杂性和不确定性,并且在大比例尺地理空间,人们经常遇到不完全、不确定的空间信息,因而普遍使用定性方法理解、分析、描述和对空间环境下结论。当所得到的信息为定性信息,缺少更为精确的地理信息的情况下,定性推理成为推导隐含信息和未知信息的唯一可行方法。定性推理可以处理非精确数据,虽然可能只提供近似的、粗略的解、丢失一些精度,但简化了推理过程,并足以解决特定的地理问题(曹菡,2002)。

§1.3 本书的组织与安排

全书系统地研究了三维空间关系:拓扑关系、方向关系、定性距离和位置关系的描述、表达及其定性推理。研究内容分七章。

第 1 章综合阐述和分析目前空间关系的研究现状。本章从二维空间和三维空间两个范围介绍国内外拓扑关系、方向关系和距离关系的研究现状。

第 2 章研究三维空间拓扑关系的描述模型及其定性推理。采用 SDS 数据模型,研究三维空间点(0-单纯形)、线(1-单纯形)、面(2-单纯形)、体(3-单纯形)间的拓扑关系,以及拓扑关系定性推理,其中包括点-线、点-面、点-体、线-线、线-面、线-体、面-面、面-体和体-体拓扑关系推理,用组合推理表给出推理结果。

第 3 章研究三维空间方向关系的描述模型及其定性推理。采用 SDS 数据模型,以体(3-单纯形)为参照物研究三维空间方向关系的描述模型,方向关系推理主要研究单方向关系推理,采用组合推理表给出推理结果。

第 4 章研究三维空间定性距离的描述模型及其定性推理。研究三维空间定性距离的描述方法,定性距离推理主要研究同向和反向两种定性距离推理,并给出定性距离的组合推理表。

第 5 章研究混合空间关系的定性推理。空间关系表达了空间数据间的一种约束,体现了不同空间关系之间的一种层次关系,并且不同空间关系之间不是完全相互独立的,它们之间存在着一定的联系。本章重点研究拓扑关系和方向关系间的相互关系,其中包括由拓扑关系推导方向关系,以及由方向关系推导拓扑关系,并用组合表给出推理结果。

第 6 章研究三维综合拓扑关系和方向关系的描述模型及其综合空间关系推理。分析方向区域的划分特点以及拓扑关系的特点,给出拓扑方向关系(拓扑关系+方向关系)的描述模型,根据拓扑关系的不同,给出拓扑方向关系组合推理表,即根据 B、A 间的拓扑和方向关系,以及 C、B 间的拓扑和方向关系,推导 C、A 间的拓扑和方向关系,并用组合表给出推理结果。

第 7 章研究三维空间位置关系定性描述模型及其定性推理。研究目标对象和参照物间位置关系的定性描述模型,即综合距离关系和方向关系的描述模型,并根据位置关系的定性描述模型研究位置关系的定性推理。

第 2 章　拓扑关系描述及其定性推理

空间拓扑关系是研究几何图形的拓扑性质。拓扑性质就是几何图形在作拓扑变换时保持不变的性质。拓扑性质体现的是图形整体结构上的特性，可以随意地把图形作变形（如挤压、拉伸或扭曲等），只要不把它撕裂、不发生粘连、不破坏其整体结构，拓扑性质将保持不变，因此拓扑关系是非常重要的空间关系。根据研究范围，拓扑关系可分为二维拓扑关系和三维拓扑关系。按照研究内容，拓扑关系可分为拓扑关系的描述和定性推理。目前对拓扑关系的研究主要集中于二维空间拓扑关系的描述和推理。对三维空间拓扑关系的研究才刚刚开始，缺乏采用一种便于计算机操作的数据模型以及系统研究空间物体间的拓扑关系。

本章采用单纯形数据模型，用九交矩阵模型研究三维空间点（0-单纯形）、线（1-单纯形）、面（2-单纯形）和体（3-单纯形）间的拓扑关系。根据点集理论，运用集合关系的传递性，研究三维空间二值拓扑关系的定性推理。

§2.1　拓扑关系描述

拓扑关系是在旋转、缩放等拓扑变化下保持连通性、分离性不变的空间关系（杜晓初，2005）。在 2D GIS 中，拓扑关系的描述方法主要有：点集拓扑法、DEM 方法、最小边界矩形法、区域连接法、CBM 方法、2D-STRING 法、Voronoi 图法和广义交模型描述法等，本章采用点集拓扑法描述拓扑关系。

点集拓扑学研究空间结构及空间图形在连续形变下保持不变的性质。点集拓扑法描述拓扑关系常用两种模型：四交模型和九交模型。根据点集拓扑理论，将研究对象 A 所在的空间分为三个部分：内部（$A°$）、外部（A^-）和边界（∂A），这三个部分相互排斥并构成了整个三维空间，三个部分满足下面的条件：

(1) 三个部分之间的并构成整个三维空间，即 $A° \cup \partial A \cup A^- = \mathbf{R}^3$；

(2) 三个部分相互之间的交为空集，即 $(A° \cap \partial A) \cup (A° \cap A^-) \cup (\partial A \cap A^-) = \varnothing$。

在研究两个对象 A、B 之间的拓扑关系时，如果只用两物体的内部（$A°$、$B°$）和边界（∂A、∂B）间的交集所组成的 2×2 矩阵

$$\begin{bmatrix} \partial A \cap \partial B & \partial A \cap B° \\ A° \cap \partial B & A° \cap B° \end{bmatrix} \tag{2.1}$$

来描述两物体间的拓扑关系,称这种拓扑关系描述方法为四交模型,所得 2×2 矩阵为四交矩阵(4-intersection matirx,4IM)(Egenhofer,1993)。对于精确物体,通常用空集(0)或非空集(1)表示。因此,4IM 能够区分 $2^4=16$ 种拓扑关系。当用两物体的内部(A°、B°)、边界(∂A、∂B)和外部(A^-、B^-)间的交所组成的 3×3 矩阵

$$\begin{bmatrix} \partial A\cap\partial B & \partial A\cap B^{\circ} & \partial A\cap B^- \\ A^{\circ}\cap\partial B & A^{\circ}\cap B^{\circ} & A^{\circ}\cap B^- \\ A^-\cap\partial B & A^-\cap B^{\circ} & A^-\cap B^- \end{bmatrix} \quad (2.2)$$

来描述两物体间的拓扑关系时,称这种拓扑关系描述方法为九交模型,所得 3×3 矩阵为九交矩阵(9-intersection matrix,9IM)(Egenhofer et al,1991b)。对于精确物体,九个交集的值通常用空集(0)或非空集(1)表示。理论上,9IM 能够区分 $2^9=512$ 种拓扑关系,但实际上只有很少的一部分能够实现。

在 3D-GIS 中的单纯形分别为 0-单纯形、1-单纯形、2-单纯形、3-单纯形,其对应的几何图形分别为点、线段、三角形、四面体。下面简称线段、三角形、四面体分别为:线、面、体。点是最简单的几何体,其内部只有一个点,边界为空集,因此点-点间的拓扑关系只有分离和同一两种拓扑关系。点与线、面、体间的拓扑关系只有属于和分离两种拓扑关系,拓扑关系相对简单,因此本书重点研究线、面和体间的拓扑关系:线-线、线-面、线-体、面-面、面-体、体-体。采用九交模型式(2.2)描述三维空间中的拓扑关系,用 0 表示交集为空,用 1 表示交集为非空,所得的 9IM 为 0-1 矩阵。由于式(2.2)是非对称矩阵,因此用

$$top(A,B)=\begin{bmatrix} A^{\circ}\cap B^{\circ} & A^{\circ}\cap\partial B & A^{\circ}\cap B^- \\ \partial A\cap B^{\circ} & \partial A\cap\partial B & \partial A\cap B^- \\ A^-\cap B^{\circ} & A^-\cap\partial B & A^-\cap B^- \end{bmatrix} \quad (2.3)$$

描述空间对象 A 与 B 的拓扑关系,用

$$top(B,A)=\begin{bmatrix} B^{\circ}\cap A^{\circ} & B^{\circ}\cap\partial A & B^{\circ}\cap A^- \\ \partial B\cap A^{\circ} & \partial B\cap\partial A & \partial B\cap A^- \\ B^-\cap A^{\circ} & B^-\cap\partial A & B^-\cap A^- \end{bmatrix} \quad (2.4)$$

描述 B 与 A 的拓扑关系。

1. 线-线拓扑关系

设 A 为粗线,B 为细线,则用 9IM 可以区分线-线的拓扑关系矩阵 $top(A,B)$ 如表 2.1 所示。

2. 线-面拓扑关系

设 A 为线段,B 为三角形,用 9IM 描述 A、B 间的拓扑关系如表 2.2 所示。

表 2.1 线-线拓扑关系

图形	A ——— B ———	A B ————————	A B ————————	A B ————————
$top(A,B)$	$\begin{bmatrix} 0 & 0 & 1 \\ 0 & 0 & 1 \\ 1 & 1 & 1 \end{bmatrix}$	$\begin{bmatrix} 0 & 0 & 1 \\ 0 & 1 & 1 \\ 1 & 1 & 1 \end{bmatrix}$	$\begin{bmatrix} 1 & 1 & 1 \\ 1 & 0 & 1 \\ 1 & 1 & 1 \end{bmatrix}$	$\begin{bmatrix} 1 & 0 & 0 \\ 1 & 1 & 0 \\ 1 & 1 & 1 \end{bmatrix}$
LL序号	LL1	LL2	LL3	LL4
图形	—A— —B—	A B ————	A —B—	A —B—
$top(A,B)$	$\begin{bmatrix} 1 & 0 & 0 \\ 1 & 0 & 0 \\ 1 & 1 & 1 \end{bmatrix}$	$\begin{bmatrix} 1 & 0 & 0 \\ 0 & 1 & 0 \\ 0 & 0 & 1 \end{bmatrix}$	$\begin{bmatrix} 1 & 1 & 1 \\ 0 & 1 & 1 \\ 0 & 0 & 1 \end{bmatrix}$	$\begin{bmatrix} 1 & 1 & 1 \\ 0 & 0 & 1 \\ 0 & 0 & 1 \end{bmatrix}$
LL序号	LL5	LL6	LL7	LL8
图形	B ┼ A	B│A	A │B	
$top(A,B)$	$\begin{bmatrix} 1 & 0 & 1 \\ 0 & 0 & 1 \\ 1 & 1 & 1 \end{bmatrix}$	$\begin{bmatrix} 0 & 0 & 1 \\ 1 & 0 & 1 \\ 1 & 1 & 1 \end{bmatrix}$	$\begin{bmatrix} 0 & 1 & 1 \\ 0 & 0 & 1 \\ 1 & 1 & 1 \end{bmatrix}$	
LL序号	LL9	LL10	LL11	

表 2.2 线-面拓扑关系

图形	△	△ 同一平面	△ 同一平面	△ 同一平面	△	△ 同一平面	△ 同一平面
$top(A,B)$	$\begin{bmatrix} 0 & 0 & 1 \\ 0 & 0 & 1 \\ 1 & 1 & 1 \end{bmatrix}$	$\begin{bmatrix} 0 & 0 & 1 \\ 0 & 1 & 1 \\ 1 & 1 & 1 \end{bmatrix}$	$\begin{bmatrix} 0 & 1 & 1 \\ 0 & 1 & 1 \\ 1 & 1 & 1 \end{bmatrix}$	$\begin{bmatrix} 0 & 1 & 0 \\ 0 & 1 & 0 \\ 1 & 1 & 1 \end{bmatrix}$	$\begin{bmatrix} 0 & 1 & 1 \\ 0 & 0 & 1 \\ 1 & 1 & 1 \end{bmatrix}$	$\begin{bmatrix} 1 & 1 & 1 \\ 1 & 0 & 1 \\ 1 & 1 & 1 \end{bmatrix}$	$\begin{bmatrix} 1 & 0 & 0 \\ 1 & 1 & 0 \\ 1 & 1 & 1 \end{bmatrix}$
LP序号	LP1	LP2	LP3	LP4	LP5	LP6	LP7
$top(B,A)$	$\begin{bmatrix} 0 & 0 & 1 \\ 0 & 0 & 1 \\ 1 & 1 & 1 \end{bmatrix}$	$\begin{bmatrix} 0 & 0 & 1 \\ 0 & 1 & 1 \\ 1 & 1 & 1 \end{bmatrix}$	$\begin{bmatrix} 0 & 0 & 1 \\ 1 & 1 & 1 \\ 1 & 1 & 1 \end{bmatrix}$	$\begin{bmatrix} 0 & 0 & 1 \\ 1 & 1 & 1 \\ 0 & 0 & 1 \end{bmatrix}$	$\begin{bmatrix} 0 & 0 & 1 \\ 1 & 0 & 1 \\ 1 & 1 & 1 \end{bmatrix}$	$\begin{bmatrix} 1 & 1 & 1 \\ 1 & 0 & 1 \\ 1 & 1 & 1 \end{bmatrix}$	$\begin{bmatrix} 1 & 1 & 1 \\ 0 & 1 & 1 \\ 0 & 0 & 1 \end{bmatrix}$
PL序号	PL1	PL2	PL3	PL4	PL5	PL6	PL7

第 2 章 拓扑关系描述及其定性推理

续表

图形	同一平面	同一平面	同一平面	同一平面		
$top(A,B)$	$\begin{bmatrix}1&0&0\\1&0&0\\1&1&1\end{bmatrix}$	$\begin{bmatrix}1&0&0\\0&1&0\\1&1&1\end{bmatrix}$	$\begin{bmatrix}1&1&1\\0&0&1\\1&1&1\end{bmatrix}$	$\begin{bmatrix}1&1&1\\0&1&1\\1&1&1\end{bmatrix}$	$\begin{bmatrix}0&0&1\\1&0&1\\1&1&1\end{bmatrix}$	$\begin{bmatrix}1&0&1\\0&0&1\\1&1&1\end{bmatrix}$
LP序号	LP8	LP9	LP10	LP11	LP12	LP13
$top(B,A)$	$\begin{bmatrix}1&1&1\\0&0&1\\0&0&1\end{bmatrix}$	$\begin{bmatrix}1&0&1\\0&1&1\\0&0&1\end{bmatrix}$	$\begin{bmatrix}1&0&1\\1&0&1\\1&1&1\end{bmatrix}$	$\begin{bmatrix}1&0&1\\1&1&1\\1&1&1\end{bmatrix}$	$\begin{bmatrix}0&1&1\\0&0&1\\1&1&1\end{bmatrix}$	$\begin{bmatrix}1&0&1\\0&0&1\\1&1&1\end{bmatrix}$
PL序号	PL8	PL9	PL10	PL11	PL12	PL13

3. 线-体间的拓扑关系

线-体拓扑关系是 1-单纯形(线段)和 3-单纯形(四面体)间的拓扑关系。体-线拓扑关系是四面体与线段间的拓扑关系。设 A 为四面体,B 为线段,用 9IM 能够区分的拓扑关系如表 2.3 所示。

表 2.3 线-体拓扑关系

图形						
$top(A,B)$	$\begin{bmatrix}0&0&1\\0&0&1\\1&1&1\end{bmatrix}$	$\begin{bmatrix}0&0&1\\0&1&1\\1&1&1\end{bmatrix}$	$\begin{bmatrix}0&0&1\\1&0&1\\1&1&1\end{bmatrix}$	$\begin{bmatrix}0&0&1\\1&1&1\\1&1&1\end{bmatrix}$	$\begin{bmatrix}0&0&1\\1&1&1\\0&0&1\end{bmatrix}$	$\begin{bmatrix}1&1&1\\1&0&1\\1&1&1\end{bmatrix}$
TL序号	TL1	TL2	TL3	TL4	TL5	TL6
$top(B,A)$	$\begin{bmatrix}0&0&1\\0&0&1\\1&1&1\end{bmatrix}$	$\begin{bmatrix}0&0&1\\0&1&1\\1&1&1\end{bmatrix}$	$\begin{bmatrix}0&1&1\\0&0&1\\1&1&1\end{bmatrix}$	$\begin{bmatrix}0&1&1\\0&1&1\\1&1&1\end{bmatrix}$	$\begin{bmatrix}0&1&0\\0&1&0\\1&1&1\end{bmatrix}$	$\begin{bmatrix}1&1&1\\1&0&1\\1&1&1\end{bmatrix}$
LT序号	LT1	LT2	LT3	LT4	LT5	LT6

图形						
$top(A,B)$	$\begin{bmatrix}1&0&1\\1&1&1\\1&1&1\end{bmatrix}$	$\begin{bmatrix}1&0&1\\1&0&1\\1&1&1\end{bmatrix}$	$\begin{bmatrix}1&0&1\\0&1&1\\0&0&1\end{bmatrix}$	$\begin{bmatrix}1&1&1\\0&1&1\\0&0&1\end{bmatrix}$	$\begin{bmatrix}1&1&1\\0&0&1\\0&0&1\end{bmatrix}$	
TL序号	TL7	TL8	TL9	TL10	TL11	
$top(B,A)$	$\begin{bmatrix}1&1&1\\0&1&1\\1&1&1\end{bmatrix}$	$\begin{bmatrix}1&1&1\\0&0&1\\1&1&1\end{bmatrix}$	$\begin{bmatrix}1&0&0\\0&1&0\\1&1&1\end{bmatrix}$	$\begin{bmatrix}1&0&0\\1&1&0\\1&1&1\end{bmatrix}$	$\begin{bmatrix}1&0&0\\1&0&0\\1&1&1\end{bmatrix}$	
LT序号	LT7	LT8	LT9	LT10	LT11	

4. 面-面拓扑关系

设 A 为细三角形，B 为粗三角形，用 9IM 可以区分的 A 与 B 间的拓扑关系 $top(A,B)$ 如表 2.4 所示。

表 2.4 面-面拓扑关系

图形						
$top(A,B)$	$\begin{bmatrix}0&0&1\\0&0&1\\1&1&1\end{bmatrix}$	$\begin{bmatrix}0&0&1\\0&1&1\\1&1&1\end{bmatrix}$	$\begin{bmatrix}1&1&1\\1&1&1\\1&1&1\end{bmatrix}$	$\begin{bmatrix}1&0&0\\1&1&0\\1&1&1\end{bmatrix}$	$\begin{bmatrix}1&0&0\\1&0&0\\1&1&1\end{bmatrix}$	$\begin{bmatrix}1&1&1\\0&1&1\\0&0&1\end{bmatrix}$
序号	PP1	PP2	PP3	PP4	PP5	PP6
图形						
$top(A,B)$	$\begin{bmatrix}1&1&1\\0&0&1\\0&0&1\end{bmatrix}$	$\begin{bmatrix}1&0&0\\0&1&0\\1&1&1\end{bmatrix}$	$\begin{bmatrix}0&1&1\\0&1&1\\1&1&1\end{bmatrix}$	$\begin{bmatrix}0&1&1\\0&0&1\\1&1&1\end{bmatrix}$	$\begin{bmatrix}1&1&1\\1&0&1\\1&1&1\end{bmatrix}$	$\begin{bmatrix}1&1&1\\0&1&1\\1&1&1\end{bmatrix}$
序号	PP7	PP8	PP9	PP10	PP11	PP12
图形						
$top(A,B)$	$\begin{bmatrix}1&1&1\\0&0&1\\1&1&1\end{bmatrix}$	$\begin{bmatrix}1&0&1\\0&1&1\\1&1&1\end{bmatrix}$	$\begin{bmatrix}0&0&1\\1&1&1\\1&1&1\end{bmatrix}$	$\begin{bmatrix}0&0&1\\1&0&1\\1&1&1\end{bmatrix}$	$\begin{bmatrix}1&0&1\\1&1&1\\1&1&1\end{bmatrix}$	$\begin{bmatrix}1&0&1\\1&0&1\\1&1&1\end{bmatrix}$
序号	PP13	PP14	PP15	PP16	PP17	PP18

在 3D GIS 中，用 9IM 不仅可以区分两个 2-单纯形在同一平面上的 8 种拓扑关系，还可以区分当两个单纯形不在同一平面上时的 10 种拓扑关系。因此，在 3D GIS 中，9IM 可以区分 18 种面-面拓扑关系。

5. 面-体拓扑关系

面-体拓扑关系是面与四面体间的拓扑关系。设 A 为四面体，B 为面，用 9IM 可以区分的 A 与 B 间的拓扑关系如表 2.5 所示。

表 2.5 面-体拓扑关系

图形					
$top(A,B)$	$\begin{bmatrix}0&0&1\\0&0&1\\1&1&1\end{bmatrix}$	$\begin{bmatrix}0&0&1\\0&1&1\\1&1&1\end{bmatrix}$	$\begin{bmatrix}0&0&1\\1&0&1\\1&1&1\end{bmatrix}$	$\begin{bmatrix}0&0&1\\1&1&1\\1&1&1\end{bmatrix}$	$\begin{bmatrix}0&0&1\\1&1&1\\0&0&1\end{bmatrix}$
序号	TP1	TP2	TP3	TP4	TP5
$top(B,A)$	$\begin{bmatrix}0&0&1\\0&0&1\\1&1&1\end{bmatrix}$	$\begin{bmatrix}0&0&1\\0&1&1\\1&1&1\end{bmatrix}$	$\begin{bmatrix}0&1&1\\0&0&1\\1&1&1\end{bmatrix}$	$\begin{bmatrix}0&1&1\\0&1&1\\1&1&1\end{bmatrix}$	$\begin{bmatrix}0&1&0\\0&1&0\\1&1&1\end{bmatrix}$
序号	PT1	PT2	PT3	PT4	PT5

图形					
$top(A,B)$	$\begin{bmatrix}1&1&1\\1&1&1\\1&1&1\end{bmatrix}$	$\begin{bmatrix}1&1&1\\0&1&1\\0&0&1\end{bmatrix}$	$\begin{bmatrix}1&1&1\\0&0&1\\0&0&1\end{bmatrix}$	$\begin{bmatrix}1&0&1\\1&0&1\\1&1&1\end{bmatrix}$	
序号	TP6	TP7	TP8	TP9	
$top(B,A)$	$\begin{bmatrix}1&1&1\\1&1&1\\1&1&1\end{bmatrix}$	$\begin{bmatrix}1&0&0\\1&1&0\\1&1&1\end{bmatrix}$	$\begin{bmatrix}1&0&0\\1&0&0\\1&1&1\end{bmatrix}$	$\begin{bmatrix}1&1&1\\0&0&1\\1&1&1\end{bmatrix}$	
序号	PT6	PT7	PT8	PT9	

6. 体-体拓扑关系

体-体拓扑关系是四面体与四面体间的拓扑关系。设 A 为四面体(细体),B 为四面体(粗体),用 9IM 可以区分的 A 与 B 间的拓扑关系 $top(A,B)$ 如表 2.6 所示。

表 2.6 体-体拓扑关系

图形				
$top(A,B)$	$\begin{bmatrix}0&0&1\\0&0&1\\1&1&1\end{bmatrix}$	$\begin{bmatrix}0&0&1\\0&1&1\\1&1&1\end{bmatrix}$	$\begin{bmatrix}1&1&1\\1&1&1\\1&1&1\end{bmatrix}$	$\begin{bmatrix}1&1&1\\0&1&1\\0&0&1\end{bmatrix}$
关系名称	disjoint	meet	overlap	cover
图形				
$top(A,B)$	$\begin{bmatrix}1&1&1\\0&0&1\\0&0&1\end{bmatrix}$	$\begin{bmatrix}1&0&0\\1&0&0\\1&1&1\end{bmatrix}$	$\begin{bmatrix}1&0&0\\1&1&0\\1&1&1\end{bmatrix}$	$\begin{bmatrix}1&0&0\\0&1&0\\0&0&1\end{bmatrix}$
关系名称	contain	inside	coveredby	equal

根据表 2.1 至表 2.6 可知,用九交矩阵模型可以区分 11 种线-线拓扑关系、13 种线-面拓扑关系、11 种线-体拓扑关系、18 种面-面拓扑关系、9 种面-体拓扑关系、8 种体-体拓扑关系(Liu Xin et al,2008a)。

描述空间对象 A 与 B 拓扑关系的九交矩阵 $top(A,B)$ 与描述 B 与 A 拓扑关系的九交矩阵 $top(B,A)$ 互为转置矩阵。拓扑关系集 $\{top(A,B)\}$ 描述了 A 与 B 所有可能的拓扑关系。对于两个维数相同的单纯形 A 与 B,由图形的对称性可知,$\{top(A,B)\}=\{top(B,A)\}$,如线-线、面-面和体-体之间的拓扑关系集。对维数不同的单纯形 A 与 B,$\{top(A,B)\}\neq\{top(B,A)\}$,如体-面、体-线、体-点、面-线、面-点和线-点之间的拓扑关系集。

§2.2 拓扑关系定性推理

拓扑关系推理是在推理过程中只考虑拓扑信息,不考虑距离和方位信息的推理。拓扑关系推理是利用已知的拓扑关系 $top(A,B)$ 和 $top(B,C)$,推导未知的 $top(A,C)$,即

$$top(A,B) \wedge top(B,C) \rightarrow top(A,C) \quad (2.5)$$

本书是用两个空间物体的内部、边界和外部集合间的交来描述表达两者的拓扑关系的,因此研究拓扑关系的推理问题实质上是用集合论的相关知识研究拓扑关系定性推理。根据集合论得到式(2.6)至式(2.13)(Egenhofer,1994)。

$$A \cap B = \neg \varnothing \wedge B \subseteq C \Rightarrow A \cap C = \neg \varnothing \quad (2.6)$$

证明:设任意 $x \in A \cap B$,则 $x \in A$ 且 $x \in B$,由于 $B \subseteq C$,所以 $x \in C$,因此 $x \in A \cap C$,即 $A \cap C = \neg \varnothing$。

$$A \supseteq B \wedge B \cap C \neq \neg \varnothing \Rightarrow A \cap C = \neg \varnothing \quad (2.7)$$

证明:同式(2.6)。

$$A \cap B = \varnothing \wedge B \supseteq C \Rightarrow A \cap C = \varnothing \quad (2.8)$$

证明:设任意 $x \in C$,由于 $B \supseteq C$,所以 $x \in B$。由 $A \cap B = \varnothing$ 知 $x \notin A$,因此 $A \cap C = \varnothing$。

$$A \subseteq B \wedge B \cap C = \varnothing \Rightarrow A \cap C = \varnothing \quad (2.9)$$

证明:同式(2.8)。

$$A \cap B = \neg \varnothing \wedge B \subseteq (C_0 \cup C_1) \Rightarrow A \cap (C_0 \cup C_1) = \neg \varnothing \quad (2.10)$$

证明:将 $(C_0 \cup C_1)$ 取代式(2.6)中的 C 即得。

$$(A_0 \cup A_1) \supseteq B \wedge B \cap C = \neg \varnothing \Rightarrow (A_0 \cup A_1) \cap C = \neg \varnothing \quad (2.11)$$

证明:根据式(2.10)即得。

$$A \cap (B_0 \cup B_1) = \varnothing \wedge (B_0 \cup B_1) \supseteq C \Rightarrow A \cap C = \varnothing \quad (2.12)$$

证明:$(B_0 \cup B_1)$ 取代式(2.9)中的 B 即得。

设 A、B 为三维空间中的空间物体(目标对象),$a_i, a_j, a_k \in \{A^\circ, \partial A, A^-\}$,$b_l, b_m, b_n \in \{B^\circ, \partial B, B^-\}$,且 $a_i \neq a_j \neq a_k$,$b_l \neq b_m \neq b_n$。令 $I(a_i, b_l) = a_i \cap b_l$,根据式(2.6)至式(2.12)得式(2.13)至式(2.18)。

$$a_i \subseteq b_l \Leftrightarrow I(a_i, b_l) = \neg \varnothing \wedge I(a_i, b_m) = \varnothing \wedge I(a_i, b_n) = \varnothing \quad (2.13)$$

证明:由 $b_l, b_m, b_n \in \{B^\circ, \partial B, B^-\}$,且 $b_l \neq b_m \neq b_n$ 知 $b_l \cup b_m \cup b_n = \mathbf{R}^3$,又因 $a_i \subseteq \mathbf{R}^3$,所以 $a_i \subseteq b_l \cup b_m \cup b_n$。

$$\left.\begin{array}{l} a_i \subseteq b_l \\ a_i \subseteq b_l \cup b_m \cup b_n \\ (b_l \cap b_m) \cup (b_l \cap b_n) \cup (b_m \cap b_n) = \varnothing \end{array}\right\} \Rightarrow I(a_i, b_l) = \neg \varnothing \wedge I(a_i, b_m) = \varnothing \wedge I(a_i, b_n) = \varnothing$$

$$\left.\begin{array}{l} I(a_i, b_l) = \neg \varnothing \wedge I(a_i, b_m) = \varnothing \wedge I(a_i, b_n) = \varnothing \\ (b_l \cap b_m) \cup (b_l \cap b_n) \cup (b_m \cap b_n) = \varnothing \\ a_i \subseteq b_l \cup b_m \cup b_n \end{array}\right\} \Rightarrow a_i \subseteq b_l$$

$$a_i \supseteq b_l \Leftrightarrow I(a_i,b_l) = \neg \varnothing \wedge I(a_j,b_l) = \varnothing \wedge I(a_k,b_l) = \varnothing \quad (2.14)$$

证明：同式(2.13)。

$$I(a_i,b_l) = \neg \varnothing \wedge I(a_i,b_m) = \neg \varnothing \wedge I(a_i,b_n) = \varnothing \Rightarrow a_i \subseteq (b_l \bigcup b_m) \quad (2.15)$$

证明：由 $b_l,b_m,b_n \in \{B°,\partial B,B^-\}$，且 $b_l \neq b_m \neq b_n$ 知 $b_l \bigcup b_m \bigcup b_n = \mathbf{R}^3$，又 $a_i \subseteq \mathbf{R}^3$，所以 $a_i \subseteq b_l \bigcup b_m \bigcup b_n$。

$$\left.\begin{array}{l} I(a_i,b_l) = \neg \varnothing \wedge I(a_i,b_m) = \neg \varnothing \wedge I(a_i,b_n) = \varnothing \\ a_i \subseteq (b_l \bigcup b_m \bigcup b_n) \\ (b_l \bigcap b_m) \bigcup (b_l \bigcap b_n) \bigcup (b_m \bigcap b_n) = \varnothing \end{array}\right\} \Rightarrow a_i \subseteq (b_l \bigcup b_m)$$

$$I(a_i,b_l) = \neg \varnothing \wedge I(a_j,b_l) = \neg \varnothing \wedge I(a_k,b_l) = \varnothing \Rightarrow (a_i \bigcup a_j) \Leftrightarrow b_l \quad (2.16)$$

证明：同式(2.15)。

$$\neg(I(a_i,b_l) = \varnothing \wedge I(a_i,b_m) = \varnothing) \Rightarrow a_i \bigcap (b_l \bigcup b_m) = \neg \varnothing \quad (2.17)$$

证明：$\neg(I(a_i,b_l) = \varnothing \wedge I(a_i,b_m) = \varnothing) \Rightarrow (a_i \bigcap b_l) \bigcup (a_i \bigcap b_m) = \neg \varnothing \Rightarrow a_i \bigcap (b_l \bigcup b_m) = \neg \varnothing$。

$$a_i \bigcap (b_l \bigcup b_m) = \varnothing \Leftrightarrow I(a_i,b_l) = \varnothing \wedge I(a_i,b_m) = \varnothing \quad (2.18)$$

证明：$a_i \bigcap (b_l \bigcup b_m) = \varnothing \Leftrightarrow (a_i \bigcap b_l) \bigcup (a_i \bigcap b_m) = \varnothing \Leftrightarrow I(a_i,b_l) \wedge I(a_i,b_m) = \varnothing$。

设 A、B、C 为三维空间中的空间物体，$a_i,a_j,a_k \in \{A°,\partial A,A^-\}$，$b_l,b_m,b_n \in \{B°,\partial B,B^-\}$，$c_o,c_p,c_q \in \{C°,\partial C,C^-\}$，且 $a_i \neq a_j \neq a_k$，$b_l \neq b_m \neq b_n$，$c_o \neq c_p \neq c_q$，用 $I_x[a_i,b_l]$ 表示空间物体 A 和 B 的内部、边界和外部的交；$I_y[b_l,c_o]$ 表示空间物体 B 和 C 的内部、边界和外部的交；$I_z[a_i,c_o]$ 表示空间物体 A 和 C 的内部、边界和外部的交，且交为空集用 0 表示，交非空用 1 表示。根据式(2.6)至式(2.18)得到式(2.19)至式(2.26)。

将式(2.13)用于式(2.6)，得

$$\left.\begin{array}{l} I_x[a_i,b_l] = 1 \\ I_y[b_l,c_o] = 0 \wedge I_y[b_l,c_p] = 0 \wedge I_y[b_l,c_q] = 1 \end{array}\right\} \Rightarrow I_z[a_i,c_q] = 1 \quad (2.19)$$

将式(2.14)用于式(2.7)，得

$$\left.\begin{array}{l} I_x[a_i,b_l] = 1 \wedge I_x[a_j,b_l] = 0 \wedge I_x[a_k,b_l] = 0 \\ I_y[b_l,c_o] = 1 \end{array}\right\} \Rightarrow I_z[a_i,c_o] = 1 \quad (2.20)$$

将式(2.14)用于式(2.8)，得

$$\left.\begin{array}{l} I_x[a_i,b_l] = 0 \\ I_y[b_l,c_o] = 1 \wedge I_y[b_m,c_o] = 0 \wedge I_y[b_n,c_o] = 0 \end{array}\right\} \Rightarrow I_z[a_i,c_o] = 0 \quad (2.21)$$

将式(2.13)用于式(2.9)，得

第 2 章 拓扑关系描述及其定性推理

$$\left.\begin{array}{l}I_x[a_i,b_l]=1 \wedge I_x[a_i,b_m]=0 \wedge I_x[a_i,b_n]=0 \\ I_y[b_l,c_o]=0\end{array}\right\} \Rightarrow I_z[a_i,c_o]=0 \quad (2.22)$$

将式(2.15)和式(2.16)用于式(2.10),得

$$\left.\begin{array}{l}I_x[a_i,b_l]=1 \\ I_y[b_l,c_o]=1 \wedge I_y[b_l,c_p]=0 \wedge I_y[b_l,c_q]=1\end{array}\right\} \Rightarrow \neg(I_z[a_i,c_o]=0 \wedge I_z[a_i,c_q]=0)$$
$$(2.23)$$

将式(2.16)和式(2.17)用于式(2.11),得

$$\left.\begin{array}{l}I_x[a_i,b_l]=1 \wedge I_x[a_j,b_l]=1 \wedge I_x[a_k,b_l]=0 \\ I_y[b_l,c_o]=1\end{array}\right\} \Rightarrow \neg(I_z[a_i,c_o]=0 \wedge I_z[a_j,c_o]=0)$$
$$(2.24)$$

将式(2.16)和式(2.18)用于式(2.12),得

$$\left.\begin{array}{l}I_x[a_i,b_l]=0 \wedge I_x[a_i,b_m]=0 \\ I_y[b_l,c_o]=1 \wedge I_y[b_m,c_o]=1 \wedge I_y[b_n,c_o]=0\end{array}\right\} \Rightarrow I_z[a_i,c_o]=0 \quad (2.25)$$

将式(2.15)和式(2.18)用于式(2.12),得

$$\left.\begin{array}{l}I_x[a_i,b_l]=1 \wedge I_x[a_i,b_m]=0 \wedge I_x[a_i,b_n]=1 \\ I_y[b_l,c_o]=0 \wedge I_y[b_n,c_o]=0\end{array}\right\} \Rightarrow I_z[a_i,c_o]=0 \quad (2.26)$$

根据式(2.19)至式(2.26),可以利用已知的拓扑关系矩阵 $top(A,B)$ 和 $top(B,C)$,推导出未知的 $top(A,C)$,根据 $top(A,C)$ 来研究 A 与 C 间的拓扑关系。下面举例说明如何利用式(2.19)至式(2.26)进行拓扑关系定性推理。

例 2.1 设用九交矩阵描述的空间物体 A 与 B 的拓扑关系矩阵 $top(A,B)=\begin{bmatrix}0 & 0 & 1\\ 0 & 1 & 1\\ 1 & 1 & 1\end{bmatrix}$,用九交矩阵描述的空间物体 B 与 C 的拓扑关系矩阵 $top(B,C)=\begin{bmatrix}1 & 1 & 1\\ 0 & 1 & 1\\ 0 & 0 & 1\end{bmatrix}$,二元算子 $\wedge$ 表示两个拓扑关系矩阵的组合,即根据已知的拓扑关系矩阵 $top(A,B)$ 和 $top(B,C)$,推导拓扑关系矩阵 $top(A,C)$。

解:由式(2.19)得

$$\left.\begin{array}{l}I_x[A^\circ,B^-]=1 \\ I_y[B^-,C^\circ]=0 \wedge I_y[B^-,\partial C]=0 \wedge I_y[B^-,C^-]=1\end{array}\right\} \Rightarrow I_z[A^\circ,C^-]=1$$

$$\left.\begin{array}{l}I_x[\partial A,B^-]=1 \\ I_y[B^-,C^\circ]=0 \wedge I_y[B^-,\partial C]=0 \wedge I_y[B^-,C^-]=1\end{array}\right\} \Rightarrow I_z[\partial A,C^-]=1$$

$$I_x[A^-,B^-]=1$$
$$I_y[B^-,C°]=0 \wedge I_y[B^-,\partial C]=0 \wedge I_y[B^-,C^-]=1 \bigg\} \Rightarrow I_z[A^-,C^-]=1$$

由式(2.20)得

$$I_x[A°,B°]=0 \wedge I_x[\partial A,B°]=0 \wedge I_x[A^-,B°]=1 \bigg\}$$
$$I_y[B°,C°]=1 \qquad\qquad\qquad\qquad\qquad\qquad\qquad \Rightarrow I_z[A^-,C°]=1$$

$$I_x[A°,B°]=0 \wedge I_x[\partial A,B°]=1 \wedge I_x[A^-,B°]=1 \bigg\}$$
$$I_y[B°,\partial C]=1 \qquad\qquad\qquad\qquad\qquad\qquad\qquad \Rightarrow I_z[A^-,\partial C]=1$$

$$I_x[A°,B°]=0 \wedge I_x[\partial A,B°]=1 \wedge I_x[A^-,B°]=1 \bigg\}$$
$$I_y[B°,C^-]=1 \qquad\qquad\qquad\qquad\qquad\qquad\qquad \Rightarrow I_z[A^-,C^-]=1$$

由式(2.21)得

$$I_x[A°,B°]=0$$
$$I_y[B°,C°]=1 \wedge I_y[\partial B,C°]=0 \wedge I_y[B^-,C°]=0 \bigg\} \Rightarrow I_z[A°,C°]=0$$

$$I_x[\partial A,B°]=0$$
$$I_y[B°,C°]=1 \wedge I_y[\partial B,C°]=0 \wedge I_y[B^-,C°]=0 \bigg\} \Rightarrow I_z[\partial A,C°]=0$$

由式(2.22)得

$$I_x[A°,B°]=0 \wedge I_x[A°,\partial B]=0 \wedge I_x[A°,B^-]=1 \bigg\}$$
$$I_y[B^-,C°]=0 \qquad\qquad\qquad\qquad\qquad\qquad\qquad \Rightarrow I_z[A°,C°]=0$$

$$I_x[A°,B°]=0 \wedge I_x[A°,\partial B]=0 \wedge I_x[A°,B^-]=1 \bigg\}$$
$$I_y[B^-,\partial C]=0 \qquad\qquad\qquad\qquad\qquad\qquad\qquad \Rightarrow I_z[A°,\partial C]=0$$

由式(2.23)得

$$I_x[\partial A,\partial B]=1$$
$$I_y[\partial B,C°]=0 \wedge I_y[\partial B,\partial C]=1 \wedge I_y[\partial B,C^-]=1 \bigg\} \Rightarrow \neg(I_z[\partial A,\partial C]=0 \wedge$$
$$\qquad\qquad\qquad\qquad\qquad\qquad\qquad\qquad\qquad\qquad\qquad I_z[\partial A,C^-]=0)$$

$$I_x[A^-,\partial B]=1$$
$$I_y[\partial B,C°]=0 \wedge I_y[\partial B,\partial C]=1 \wedge I_y[\partial B,C^-]=1 \bigg\} \Rightarrow \neg(I_z[A^-,\partial C]=0 \wedge$$
$$\qquad\qquad\qquad\qquad\qquad\qquad\qquad\qquad\qquad\qquad\qquad I_z[A^-,C^-]=0)$$

由式(2.24)得

$$I_x[A°,\partial B]=0 \wedge I_x[\partial A,\partial B]=1 \wedge I_x[A^-,\partial B]=1 \bigg\}$$
$$I_y[\partial B,\partial C]=1 \qquad\qquad\qquad\qquad\qquad\qquad\qquad \Rightarrow \neg(I_z[\partial A,\partial C]=0 \wedge$$
$$\qquad\qquad\qquad\qquad\qquad\qquad\qquad\qquad\qquad\qquad\qquad I_z[A^-,\partial C]=0)$$

第 2 章 拓扑关系描述及其定性推理　　29

$$\left.\begin{array}{l}I_x[A°,\partial B]=0 \wedge I_x[\partial A,\partial B]=1 \wedge I_x[A^-,\partial B]=1 \\ I_y[\partial B,C^-]=1\end{array}\right\} \Rightarrow \neg\,(I_z[\partial A,C^-]=0 \wedge \\ \hspace{7cm} I_z[A^-,C^-]=0)$$

由式(2.25)得

$$\left.\begin{array}{l}I_x[A°,B°]=0 \wedge I_x[A°,\partial B]=0 \\ I_y[B°,\partial C]=1 \wedge I_y[\partial B,\partial C]=1 \wedge I_y[B^-,\partial C]=0\end{array}\right\} \Rightarrow I_z[A°,\partial C]=0$$

由式(2.26)得

$$\left.\begin{array}{l}I_x[\partial A,B°]=0 \wedge I_x[\partial A,\partial B]=1 \wedge I_x[\partial A,B^-]=1 \\ I_y[B°,C°]=0 \wedge I_y[\partial B,C°]=0\end{array}\right\} \Rightarrow I_z[\partial A,C°]=0$$

根据以上推理得

$$\begin{bmatrix} 0 & 0 & 1 \\ 0 & 1 & 1 \\ 1 & 1 & 1 \end{bmatrix} \wedge \begin{bmatrix} 1 & 1 & 1 \\ 0 & 1 & 1 \\ 0 & 0 & 1 \end{bmatrix} \rightarrow \begin{bmatrix} 0 & 0 & 1 \\ 0 & - & 1 \\ 1 & 1 & 1 \end{bmatrix} \tag{2.27}$$

当 $\begin{bmatrix} 0 & 0 & 1 \\ 0 & 1 & 1 \\ 1 & 1 & 1 \end{bmatrix}$ 表示面-面拓扑关系，$\begin{bmatrix} 1 & 1 & 1 \\ 0 & 1 & 1 \\ 0 & 0 & 1 \end{bmatrix}$ 表示面-面拓扑关系时，定性推理的结果为

$$\begin{bmatrix} 0 & 0 & 1 \\ 0 & 0 & 1 \\ 1 & 1 & 1 \end{bmatrix} 或 \begin{bmatrix} 0 & 0 & 1 \\ 0 & 1 & 1 \\ 1 & 1 & 1 \end{bmatrix}$$

简写为 PP2∧PP6→PP1∨PP2，即当 A 与 B 间的拓扑关系为 PP2，B 与 C 间的拓扑关系为 PP6 时，A 与 C 间的拓扑关系为 PP1 或 PP2。同理，当 A 与 B 间的拓扑关系为 TL2，B 与 C 间的拓扑关系为 LL7 时，A 与 C 间的拓扑关系为 TL1 或 TL2，即 TL2∧LL7→TL1∨TL2。

例 2.2　设用九交矩阵描述的空间物体 A 与 B 的拓扑关系矩阵 $top(A,B)=$ $\begin{bmatrix} 0 & 0 & 1 \\ 1 & 0 & 1 \\ 1 & 1 & 1 \end{bmatrix}$，用九交矩阵描述的空间物体 B 与 C 的拓扑关系矩阵 $top(B,C)=$ $\begin{bmatrix} 1 & 0 & 0 \\ 0 & 1 & 0 \\ 0 & 0 & 1 \end{bmatrix}$，二元算子∧表示两个拓扑关系矩阵的组合，即根据已知拓扑关系矩阵 $top(A,B)$ 和 $top(B,C)$，推导拓扑关系矩阵 $top(A,C)$。

解：由式(2.19)得

$I_x[\partial A, B^\circ] = 1$
$I_y[B^\circ, C^\circ] = 1 \land I_y[B^\circ, \partial C] = 0 \land I_y[B^\circ, C^-] = 0 \Big\} \Rightarrow I_z[\partial A, C^\circ] = 1$

$I_x[A^-, B^\circ] = 1$
$I_y[B^\circ, C^\circ] = 1 \land I_y[B^\circ, \partial C] = 0 \land I_y[B^\circ, C^-] = 0 \Big\} \Rightarrow I_z[A^-, C^\circ] = 1$

$I_x[A^-, \partial B] = 1$
$I_y[\partial B, C^\circ] = 0 \land I_y[\partial B, \partial C] = 1 \land I_y[\partial B, C^-] = 0 \Big\} \Rightarrow I_z[A^-, \partial C] = 1$

$I_x[A^\circ, B^-] = 1$
$I_y[B^-, C^\circ] = 0 \land I_y[B^-, \partial C] = 0 \land I_y[B^-, C^-] = 1 \Big\} \Rightarrow I_z[A^\circ, C^-] = 1$

$I_x[\partial A, B^-] = 1$
$I_y[B^-, C^\circ] = 0 \land I_y[B^-, \partial C] = 0 \land I_y[B^-, C^-] = 1 \Big\} \Rightarrow I_z[\partial A, C^-] = 1$

$I_x[A^-, B^-] = 1$
$I_y[B^-, C^\circ] = 0 \land I_y[B^-, \partial C] = 0 \land I_y[B^-, C^-] = 1 \Big\} \Rightarrow I_z[A^-, C^-] = 1$

由式(2.20)得

$I_x[A^\circ, \partial B] = 0 \land I_x[\partial A, \partial B] = 0 \land I_x[A^-, \partial B] = 1$
$I_y[\partial B, \partial C] = 1$ $\Big\} \Rightarrow I_z[A^-, \partial C] = 1$

由式(2.21)得

$I_x[A^\circ, B^\circ] = 0$
$I_y[B^\circ, C^\circ] = 1 \land I_y[B^\circ, \partial C] = 0 \land I_y[B^\circ, C^-] = 0 \Big\} \Rightarrow I_z[A^\circ, C^\circ] = 0$

$I_x[\partial A, \partial B] = 0$
$I_y[\partial B, C^\circ] = 0 \land I_y[\partial B, \partial C] = 1 \land I_y[\partial B, C^-] = 0 \Big\} \Rightarrow I_z[\partial A, \partial C] = 0$

由式(2.22)得

$I_x[A^\circ, B^\circ] = 0 \land I_x[A^\circ, \partial B] = 0 \land I_x[A^\circ, B^-] = 1$
$I_y[B^-, C^\circ] = 0$ $\Big\} \Rightarrow I_z[A^\circ, C^\circ] = 0$

$I_x[A^\circ, B^\circ] = 0 \land I_x[A^\circ, \partial B] = 0 \land I_x[A^\circ, B^-] = 1$
$I_y[B^-, \partial C] = 0$ $\Big\} \Rightarrow I_z[A^\circ, \partial C] = 0$

由式(2.26)得

$I_x[\partial A, B^\circ] = 1 \land I_x[\partial A, \partial B] = 0 \land I_x[\partial A, B^-] = 1$
$I_y[B^\circ, \partial C] = 0 \land I_y[B^-, \partial C] = 0$ $\Big\} \Rightarrow I_z[\partial A, \partial C] = 0$

根据以上推理得

$$\begin{bmatrix} 0 & 0 & 1 \\ 1 & 0 & 1 \\ 1 & 1 & 1 \end{bmatrix} \land \begin{bmatrix} 1 & 0 & 0 \\ 0 & 1 & 0 \\ 0 & 0 & 1 \end{bmatrix} \rightarrow \begin{bmatrix} 0 & 0 & 1 \\ 1 & 0 & 1 \\ 1 & 1 & 1 \end{bmatrix} \quad (2.28)$$

式(2.28)表示,当 $top(A,B) = \begin{bmatrix} 0 & 0 & 1 \\ 1 & 0 & 1 \\ 1 & 1 & 1 \end{bmatrix}$, $top(B,C) = \begin{bmatrix} 1 & 0 & 0 \\ 0 & 1 & 0 \\ 0 & 0 & 1 \end{bmatrix}$ 时,A 与 C 的拓扑关系矩阵是 $\begin{bmatrix} 0 & 0 & 1 \\ 1 & 0 & 1 \\ 1 & 1 & 1 \end{bmatrix}$。定性推理的前提所描述的拓扑关系不同,定性推理的结果不同,如式(2.28)为

$$TL3 \wedge LL6 \rightarrow TL3$$
$$TP3 \wedge PP8 \rightarrow TP3$$
$$LL10 \wedge LL6 \rightarrow LL10$$

例 2.3 设用九交矩阵描述的空间物体 A 与 B 的拓扑关系矩阵 $top(A,B) = \begin{bmatrix} 1 & 1 & 1 \\ 0 & 1 & 1 \\ 0 & 0 & 1 \end{bmatrix}$,用九交矩阵描述的空间物体 B 与 C 的拓扑关系矩阵 $top(B,C) = \begin{bmatrix} 0 & 0 & 1 \\ 0 & 1 & 1 \\ 1 & 1 & 1 \end{bmatrix}$,二元算子 $\wedge$ 表示两个拓扑关系矩阵的组合,即根据两个拓扑关系矩阵 $top(A,B)$ 和 $top(B,C)$ 推导拓扑关系矩阵 $top(A,C)$。

解:由式(2.19)得

$$\left. \begin{aligned} & I_x[A^\circ, B^\circ] = 1 \\ & I_y[B^\circ, C^\circ] = 0 \wedge I_y[B^\circ, \partial C] = 0 \wedge I_y[B^\circ, C^-] = 1 \end{aligned} \right\} \Rightarrow I_z[A^\circ, C^-] = 1$$

由式(2.20)得

$$\left. \begin{aligned} & I_x[A^\circ, B^\circ] = 1 \wedge I_x[\partial A, B^\circ] = 0 \wedge I_x[A^-, B^\circ] = 0 \\ & I_y[B^\circ, C^-] = 1 \end{aligned} \right\} \Rightarrow I_z[A^\circ, C^-] = 1$$

由式(2.23)得

$$\left. \begin{aligned} & I_x[A^\circ, \partial B] = 1 \\ & I_y[\partial B, C^\circ] = 0 \wedge I_y[\partial B, \partial C] = 1 \wedge I_y[\partial B, C^-] = 1 \end{aligned} \right\} \Rightarrow \neg (I_z[A^\circ, \partial C] = 0 \wedge I_z[A^\circ, C^-] = 0)$$

$$\left. \begin{aligned} & I_x[\partial A, \partial B] = 1 \\ & I_y[\partial B, C^\circ] = 0 \wedge I_y[\partial B, \partial C] = 1 \wedge I_y[\partial B, C^-] = 1 \end{aligned} \right\} \Rightarrow \neg (I_z[\partial A, \partial C] = 0 \wedge I_z[\partial A, C^-] = 0)$$

由式(2.24)得

$$\left. \begin{aligned} & I_x[A^\circ, \partial B] = 1 \wedge I_x[\partial A, \partial B] = 1 \wedge I_x[A^-, \partial B] = 0 \\ & I_y[\partial B, \partial C] = 1 \end{aligned} \right\} \Rightarrow \neg (I_z[A^\circ, \partial C] = 0 \wedge I_z[\partial A, \partial C] = 0)$$

$$I_x[A^\circ,\partial B]=1 \wedge I_x[\partial A,\partial B]=1 \wedge I_x[A^-,\partial B]=0 \atop I_y[\partial B,C^-]=1 \Bigg\} \Rightarrow \neg(I_z[A^\circ,C^-]=0 \wedge I_z[\partial A,C^-]=0)$$

因此,根据以上推理得

$$\begin{bmatrix}1&1&1\\0&1&1\\0&0&1\end{bmatrix} \wedge \begin{bmatrix}0&0&1\\0&1&1\\1&1&1\end{bmatrix} \rightarrow \begin{bmatrix}-&-&1\\-&-&-\\-&-&-\end{bmatrix} \wedge \neg\begin{bmatrix}-&0&-\\-&0&-\\-&-&-\end{bmatrix} \wedge \neg\begin{bmatrix}-&-&-\\-&-&0\\-&-&-\end{bmatrix} \wedge$$

$$\neg\begin{bmatrix}-&0&0\\-&-&-\\-&-&-\end{bmatrix} \wedge \neg\begin{bmatrix}-&-&-\\-&0&0\\-&-&-\end{bmatrix} \qquad (2.29)$$

当推理的前提为 PP6∧PP2 时,由于在三维空间中仅有 PP1,PP2,…,PP18 共 18 种面-面拓扑关系,因此

$$PP6 \wedge PP2 \rightarrow PP2 \vee PP3 \vee PP6 \vee PP7 \vee PP9 \vee PP10 \vee PP11 \vee PP12 \vee PP13 \vee PP14 \vee PP15 \vee PP17$$

在不引起混淆的情况下,将上式简写为

$$PP6 \wedge PP2 \rightarrow PP2, PP3, PP6, PP7, PP9—PP15, PP17 \qquad (2.30)$$

同理,当 A、B 和 C 均为线段时,LL7∧LL2→LL2∨LL3∨LL7∨LL8∨LL11,简写为

$$LL7 \wedge LL2 \rightarrow LL2, LL3, LL7, LL8, LL11 \qquad (2.31)$$

同理,得

$$TP7 \wedge PP2 \rightarrow TP2, TP4—TP8 \qquad (2.32)$$

$$PP6 \wedge PT2 \rightarrow PT2—PT4, PT6, PT9 \qquad (2.33)$$

$$PL7 \wedge LT2 \rightarrow PT2—PT4, PT6, PT9 \qquad (2.34)$$

$$PP6 \wedge PL2 \rightarrow PL2—PL4, PL7—PL9, PL11, PL12 \qquad (2.35)$$

根据例 2.1 至例 2.3 推理过程可知,当 A 与 C 的拓扑关系矩阵的每个元素都确定时,A 与 C 的拓扑关系矩阵 $top(A,C)$ 具有唯一性,如例 2.2。但当 A 与 C 的拓扑关系矩阵的部分元素不确定时,有多个九交矩阵可以描述 A 与 C 的拓扑关系。因此,$top(A,C)$ 就不唯一,如例 2.1。在 A 与 C 的内部、边界和外部的九交中,只有两者边界是否相交不确定。因此,A 与 C 的拓扑关系最多有两种情况:一是两者的边界相交,在二值拓扑关系矩阵 $top(A,C)$ 中,用 1 表示;二是两者的边界不相交,在二值拓扑关系矩阵 $top(A,C)$ 中,用 0 表示。根据例 2.3 的推理结果,在 A 的内部(A°)、边界(∂A)、外部(A^-)与 C 的内部(C°)、边界(∂C)、外部(C^-)的九交中,只有 $A^\circ \cap C^- \neq \emptyset$(用 1 表示)可以确定和四个约束条件 $\neg(A^\circ \cap \partial C = \emptyset \wedge$

$\partial A \cap \partial C = \varnothing)$、$\neg (A^\circ \cap C^- = \varnothing \wedge \partial A \cap C^- = \varnothing)$、$\neg (A^\circ \cap \partial C = \varnothing \wedge A^\circ \cap C^- = \varnothing)$ 和 $\neg (\partial A \cap \partial C = \varnothing \wedge \partial A \cap C^- = \varnothing)$ 可以利用。由于 A 与 C 的几何表现形式不同，$top(A,C)$ 的取值范围不同，因此 $top(A,C)$ 也可能不同。例如，在例 2.3 中，当 A 与 C 均为体时，$top(A,C)$ 的取值范围是

$$\begin{bmatrix} 0 & 0 & 1 \\ 0 & 0 & 1 \\ 1 & 1 & 1 \end{bmatrix}, \begin{bmatrix} 0 & 0 & 1 \\ 0 & 1 & 1 \\ 1 & 1 & 1 \end{bmatrix}, \begin{bmatrix} 1 & 1 & 1 \\ 1 & 1 & 1 \\ 1 & 1 & 1 \end{bmatrix}, \begin{bmatrix} 1 & 1 & 1 \\ 0 & 1 & 1 \\ 0 & 0 & 1 \end{bmatrix}, \begin{bmatrix} 1 & 1 & 1 \\ 0 & 0 & 1 \\ 0 & 0 & 1 \end{bmatrix},$$

$$\begin{bmatrix} 1 & 0 & 0 \\ 1 & 0 & 0 \\ 1 & 1 & 1 \end{bmatrix}, \begin{bmatrix} 1 & 0 & 0 \\ 1 & 1 & 0 \\ 1 & 1 & 1 \end{bmatrix} \text{和} \begin{bmatrix} 1 & 0 & 0 \\ 0 & 1 & 0 \\ 0 & 0 & 1 \end{bmatrix}$$

当 A 与 C 均为体时，A 与 C 的拓扑关系矩阵只能是

$$\begin{bmatrix} 0 & 0 & 1 \\ 0 & 1 & 1 \\ 1 & 1 & 1 \end{bmatrix}, \begin{bmatrix} 1 & 1 & 1 \\ 1 & 1 & 1 \\ 1 & 1 & 1 \end{bmatrix}, \begin{bmatrix} 1 & 1 & 1 \\ 0 & 1 & 1 \\ 0 & 0 & 1 \end{bmatrix}, \begin{bmatrix} 1 & 1 & 1 \\ 0 & 0 & 1 \\ 0 & 0 & 1 \end{bmatrix}$$

4 个矩阵中的一个。当 A 与 C 均为面时，$top(A,C)$ 的取值范围是 PP1—PP18，$top(A,C)$ 只能是 12 个矩阵 PP2，PP3，PP6，PP7，PP9，PP10，PP11，PP12，PP13，PP14，PP15，PP17 中的一个。

根据式(2.19)至式(2.26)可以研究拓扑关系定性推理。根据 A、C 的数据类型，将拓扑关系定性推理分为：体-体、体-面、体-线、体-点、面-体、面-面、面-线、面-点、线-体、线-面、线-线、线-点、点-体、点-面、点-线、点-点拓扑关系。根据 B 的类型，将拓扑关系推理进一步划分。

1. 体-体拓扑关系推理

当 A 和 C 均为体时，推导体-体拓扑关系的方法如图 2.1 所示。

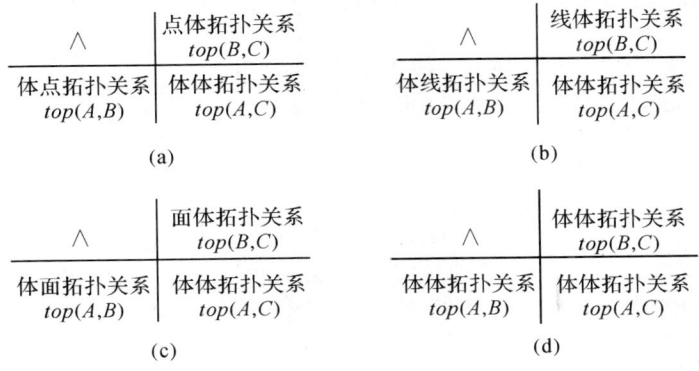

图 2.1 体-体拓扑关系推理类型

当 B 为点时,根据式(2.19)得到 A 与 C 的拓扑关系矩阵 $top(A,C)$,如表 2.7 所示。

表 2.7 当 B 为点,A 和 C 均为非点状物体时拓扑关系推理组合

$\wedge$	$B\in C^{\circ}$	$B\in \partial C$	$B\in C^{-}$
$B\in A^{\circ}$	$\begin{bmatrix} 1 & - & - \\ - & - & - \\ - & - & - \end{bmatrix}$	$\begin{bmatrix} - & 1 & - \\ - & - & - \\ - & - & - \end{bmatrix}$	$\begin{bmatrix} - & - & 1 \\ - & - & - \\ - & - & - \end{bmatrix}$
$B\in \partial A$	$\begin{bmatrix} - & - & - \\ 1 & - & - \\ - & - & - \end{bmatrix}$	$\begin{bmatrix} - & - & - \\ - & 1 & - \\ - & - & - \end{bmatrix}$	$\begin{bmatrix} - & - & - \\ - & - & 1 \\ - & - & - \end{bmatrix}$
$B\in A^{-}$	$\begin{bmatrix} - & - & - \\ - & - & - \\ 1 & - & - \end{bmatrix}$	$\begin{bmatrix} - & - & - \\ - & - & - \\ - & 1 & - \end{bmatrix}$	$\begin{bmatrix} - & - & - \\ - & - & - \\ - & - & 1 \end{bmatrix}$

当 B 为点,且 A 和 C 均为体时,根据表 2.6 和表 2.7,得到定性推理 $top(体,点)\wedge top(点,体)\to top(体,体)$,具体结果如表 2.8 所示。其中,$d=$ disjoint,$m=$ meet,$e=$ equal,$i=$ inside,$cb=$ coveredby,$ct=$ contain,$cv=$ cover,$o=$ overlap。

表 2.8 $top(体,点)\wedge top(点,体)\to top(体,体)$

$\wedge$	$B\in C^{\circ}$	$B\in \partial C$	$B\in C^{-}$
$B\in A^{\circ}$	e,i,cb,ct,cv,o	ct,cv,o	d,m,ct,cv,o
$B\in \partial A$	i,cb,o	m,e,cb,cv,o	d,m,ct,cv,o
$B\in A^{-}$	d,m,i,cb,o	d,m,i,cb,o	d,m,e,i,cb,ct,cv,o

根据式(2.19)至式(2.26),图 2.1(b)~(d)所示的三种体-体拓扑关系的定性推理结果分别如表 2.9、表 2.10 和表 2.11 所示。其中,$d=$ disjoint,$m=$ meet,$o=$ overlap,$cv=$ cover,$ct=$ contain,$cb=$ coveredby,$i=$ inside,$e=$ equal。

2. 面-面拓扑关系推理

当 A 和 C 均为面时,面-面拓扑关系推理的方法如图 2.2 所示。

$\wedge$	点面拓扑关系 $top(B,C)$
面点拓扑关系 $top(A,B)$	面面拓扑关系 $top(A,C)$

(a)

$\wedge$	线面拓扑关系 $top(B,C)$
面线拓扑关系 $top(A,B)$	面面拓扑关系 $top(A,C)$

(b)

$\wedge$	面面拓扑关系 $top(B,C)$
面面拓扑关系 $top(A,B)$	面面拓扑关系 $top(A,C)$

(c)

$\wedge$	体面拓扑关系 $top(B,C)$
面体拓扑关系 $top(A,B)$	面面拓扑关系 $top(A,C)$

(d)

图 2.2 面-面拓扑关系推理类型

根据表 2.4 和表 2.7，图 2.2(a)的定性推理结果如表 2.12 所示。根据式(2.19)至式(2.26)，图 2.2(b)～(d)所示的三种面-面拓扑关系的推理结果分别如表 2.13 至表 2.15 所示。

3. 线-线拓扑关系定性推理

当 A 为线、C 为线时，推导线-线拓扑关系的方法如图 2.3 所示。

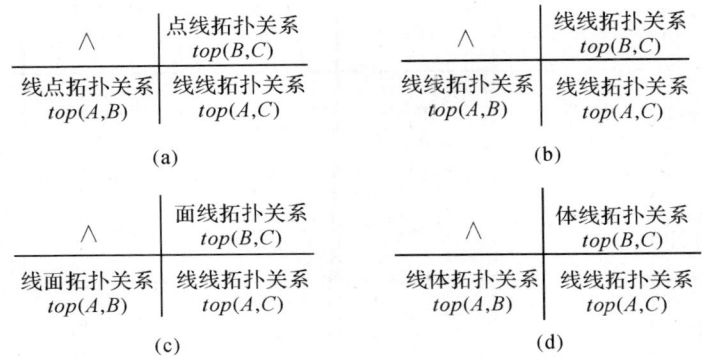

图 2.3　线-线拓扑关系推理类型

当 B 为点，且 A 和 C 均为线时，根据表 2.1 和表 2.7，得到定性推理 top(线，点)$\wedge top$(点，线)$\to top$(线，线)，具体结果见表 2.16。根据式(2.19)至式(2.26)，图2.3(b)～(d)所示的三种线-线拓扑关系推理结果分别如表 2.17 至表 2.19 所示。

4. 点-点拓扑关系定性推理

在三维空间中点-点间的拓扑关系，仅有同一和分离两种。当 B 为非点状物体（体、面或线），且 A 和 C 均为点时，定性推理的结果如表 2.20 所示。其中，$d=$ disjoint，$e=$ equal。

表 2.9 $top(体,线) \wedge top(线,体)$

$\wedge$	$\begin{bmatrix} 0 & 0 & 1 \\ 0 & 0 & 1 \\ 1 & 1 & 1 \end{bmatrix}$ LT1	$\begin{bmatrix} 0 & 0 & 1 \\ 0 & 1 & 1 \\ 1 & 1 & 1 \end{bmatrix}$ LT2	$\begin{bmatrix} 0 & 1 & 1 \\ 0 & 0 & 1 \\ 1 & 1 & 1 \end{bmatrix}$ LT3	$\begin{bmatrix} 0 & 1 & 1 \\ 0 & 1 & 1 \\ 1 & 1 & 1 \end{bmatrix}$ LT4	$\begin{bmatrix} 0 & 1 & 0 \\ 0 & 1 & 0 \\ 1 & 1 & 1 \end{bmatrix}$ LT5
$\begin{bmatrix} 0 & 0 & 1 \\ 0 & 0 & 1 \\ 1 & 1 & 1 \end{bmatrix}$ TL1	$d,m,o,cv,$ ct,cb,i,e	$d,m,o,$ cb,i	$d,m,o,$ cb,i	$d,m,o,$ cb,i	$d,m,o,$ cb,i
$\begin{bmatrix} 0 & 0 & 1 \\ 0 & 1 & 1 \\ 1 & 1 & 1 \end{bmatrix}$ TL2	$d,ct,m,$ cv,o	$d,m,e,$ cb,cv,o	d,m,o	$d,m,o,$ cb	m,cb,o
$\begin{bmatrix} 0 & 0 & 1 \\ 1 & 0 & 1 \\ 1 & 1 & 1 \end{bmatrix}$ TL3	$d,m,o,$ cv,ct	d,m,o	$d,m,e,cv,$ o,cb	d,m,o,cb	m,cb,o
$\begin{bmatrix} 0 & 0 & 1 \\ 1 & 1 & 1 \\ 1 & 1 & 1 \end{bmatrix}$ PL4	$d,m,o,$ cv,ct	d,m,cv,e	d,m,cv,o	$d,m,e,$ cb,cv,o	m,cb,o
$\begin{bmatrix} 0 & 0 & 1 \\ 1 & 1 & 1 \\ 0 & 0 & 1 \end{bmatrix}$ TL5	$d,m,o,$ cv,ct	m,o,cv	m,o,cv	m,o,cv	$m,o,cv,$ cb,e
$\begin{bmatrix} 1 & 1 & 1 \\ 1 & 0 & 1 \\ 1 & 1 & 1 \end{bmatrix}$ TL6	$d,m,o,$ cv,ct	$d,m,o,$ cv,ct	$d,m,o,$ cv,ct	$d,m,o,$ cv,ct	o
$\begin{bmatrix} 1 & 0 & 1 \\ 1 & 1 & 1 \\ 1 & 1 & 1 \end{bmatrix}$ TL7	$d,m,o,$ cv,ct	$d,m,o,$ cv	$d,m,o,$ cv,ct	d,m,o,cv	o
$\begin{bmatrix} 1 & 0 & 1 \\ 1 & 0 & 1 \\ 1 & 1 & 1 \end{bmatrix}$ TL8	$d,m,o,$ cv,ct	d,m,o	$d,m,o,$ cv,ct	d,m,o	o
$\begin{bmatrix} 1 & 0 & 1 \\ 0 & 1 & 1 \\ 0 & 0 & 1 \end{bmatrix}$ TL9	$d,m,o,$ cv,ct	m,o,cv	cv,ct,o	o,cv	o,cv
$\begin{bmatrix} 1 & 1 & 1 \\ 0 & 1 & 1 \\ 0 & 0 & 1 \end{bmatrix}$ TL10	$d,m,o,$ cv,ct	m,o,cv,ct	cv,ct,o	cv,ct,o	o,cv
$\begin{bmatrix} 1 & 1 & 1 \\ 0 & 0 & 1 \\ 0 & 0 & 1 \end{bmatrix}$ TL11	$d,m,o,$ cv,ct	ct,cv,o	cv,ct,o	cv,ct,o	cv,ct,o

→top(体,体)

$\begin{bmatrix}1&1&1\\1&0&1\\1&1&1\end{bmatrix}$	$\begin{bmatrix}1&1&1\\0&1&1\\1&1&1\end{bmatrix}$	$\begin{bmatrix}1&1&1\\0&0&1\\1&1&1\end{bmatrix}$	$\begin{bmatrix}1&0&0\\0&1&0\\1&1&1\end{bmatrix}$	$\begin{bmatrix}1&0&0\\1&1&0\\1&1&1\end{bmatrix}$	$\begin{bmatrix}1&0&0\\1&0&0\\1&1&1\end{bmatrix}$
LT6	LT7	LT8	LT9	LT10	LT11
$d,m,o,$ cb,i	$d,m,o,$ cb,i	$d,m,o,$ cb,i	$d,m,o,$ cb,i	$d,m,o,$ cb,i	$d,m,o,$ cb,i
$m,i,$ cb,o	$d,m,o,$ cb	d,m,o	m,o,cb	$m,o,$ cb,i	i,cb,o
$d,m,o,$ cb,i	$d,m,o,$ cb,i	$d,m,o,$ cb,i	o,cb,i	o,cb,i	o,cb,i
$d,m,o,$ cb,i	$d,m,o,$ cb	d,m,o	o,cb	o,cb,i	o,cb,i
o	o	o	o,cb	cb,o	o,cb,i
$d,m,o,cv,$ ct,cb,i,e	$d,m,o,$ cv,ct	$d,m,o,$ cv,ct	o	o,cb,i	o,cb,i
$d,m,o,$ cb,i	$d,m,o,cv,$ cb,e	$d,m,o,$ cv,ct	o,cb	o,cb,i	o,cb,i
$d,m,o,$ cb,i	$d,m,o,$ cb,i	$d,m,o,cv,$ ct,cb,i,e	o,cb,i	o,cb,i	o,cb,i
o	o,cv	ct,cv,o	$o,cv,$ cb,e	o,cb	o,cb,i
cv,ct,o	o,cv,ct	o,cv,ct	o,cv	o,cv	o,cb,i
cv,ct,o	o,cv,ct	o,cv,ct	o,cv,ct	o,cv,ct	$o,cv,ct,$ cb,i,e

表 2.10 $top(体,面) \wedge top(面,体)$

$\wedge$	$\begin{bmatrix} 0 & 0 & 1 \\ 0 & 0 & 1 \\ 1 & 1 & 1 \end{bmatrix}$ PT1	$\begin{bmatrix} 0 & 0 & 1 \\ 0 & 1 & 1 \\ 1 & 1 & 1 \end{bmatrix}$ PT2	$\begin{bmatrix} 0 & 1 & 1 \\ 0 & 0 & 1 \\ 1 & 1 & 1 \end{bmatrix}$ PT3	$\begin{bmatrix} 0 & 1 & 1 \\ 0 & 1 & 1 \\ 1 & 1 & 1 \end{bmatrix}$ PT4
$\begin{bmatrix} 0 & 0 & 1 \\ 0 & 0 & 1 \\ 1 & 1 & 1 \end{bmatrix}$ TP1	$d,m,e,i,$ cb,ct,cv,o	$d,i,m,$ cb,o	$d,i,m,$ cb,o	$d,i,m,$ cb,o
$\begin{bmatrix} 0 & 0 & 1 \\ 0 & 1 & 1 \\ 1 & 1 & 1 \end{bmatrix}$ TP2	$d,m,ct,$ cv,o	$d,m,e,$ cb,cv,o	d,m,o	$d,m,o,$ cb
$\begin{bmatrix} 0 & 0 & 1 \\ 1 & 0 & 1 \\ 1 & 1 & 1 \end{bmatrix}$ TP3	$d,m,ct,$ cv,o	d,m,o	$d,m,e,cv,$ cb,o	$d,m,$ cb,o
$\begin{bmatrix} 0 & 0 & 1 \\ 1 & 1 & 1 \\ 1 & 1 & 1 \end{bmatrix}$ TP4	$d,m,ct,$ cv,o	$d,m,$ cv,o	d,m,cv,o	$d,m,e,$ cb,cv,o
$\begin{bmatrix} 0 & 0 & 1 \\ 1 & 1 & 1 \\ 0 & 0 & 1 \end{bmatrix}$ TP5	$d,m,ct,$ cv,o	m,cv,o	m,cv,o	m,cv,o
$\begin{bmatrix} 1 & 1 & 1 \\ 1 & 1 & 1 \\ 1 & 1 & 1 \end{bmatrix}$ TP6	$d,m,ct,$ cv,o	$d,m,ct,$ cv,o	$d,m,ct,$ cv,o	$d,m,ct,$ cv,o
$\begin{bmatrix} 1 & 1 & 1 \\ 0 & 1 & 1 \\ 0 & 0 & 1 \end{bmatrix}$ TP7	$d,m,ct,$ cv,o	$m,ct,$ cv,o	ct,cv,o	ct,cv,o
$\begin{bmatrix} 1 & 1 & 1 \\ 0 & 0 & 1 \\ 0 & 0 & 1 \end{bmatrix}$ TP8	$d,m,ct,$ cv,o	ct,cv,o	ct,cv,o	ct,cv,o
$\begin{bmatrix} 1 & 0 & 1 \\ 1 & 0 & 1 \\ 1 & 1 & 1 \end{bmatrix}$ TP9	$d,m,ct,$ cv,o	d,m,o	$d,m,ct,$ cv,o	d,m,o

→*top*(体,体)

$\begin{bmatrix} 0 & 1 & 0 \\ 0 & 1 & 0 \\ 1 & 1 & 1 \end{bmatrix}$	$\begin{bmatrix} 1 & 1 & 1 \\ 1 & 1 & 1 \\ 1 & 1 & 1 \end{bmatrix}$	$\begin{bmatrix} 1 & 0 & 0 \\ 1 & 1 & 0 \\ 1 & 1 & 1 \end{bmatrix}$	$\begin{bmatrix} 1 & 0 & 0 \\ 1 & 0 & 0 \\ 1 & 1 & 1 \end{bmatrix}$	$\begin{bmatrix} 1 & 1 & 1 \\ 0 & 0 & 1 \\ 1 & 1 & 1 \end{bmatrix}$
PT5	PT6	PT7	PT8	PT9
$d,i,m,$ cb,o	$d,i,m,$ cb,o	$d,i,m,$ cb,o	$d,i,m,$ cb,o	$d,i,m,$ cb,o
m,cb,o	$d,i,m,$ cb,o	$i,m,$ cb,o	i,cb,o	d,m,o
m,cb,o	$d,i,m,$ cb,o	i,cb,o	i,cb,o	$d,i,m,$ cb,o
m,cb,o	$d,i,m,$ cb,o	i,cb,o	i,cb,o	d,m,o
$m,e,cb,$ cv,o	o	cb,o	i,cb,o	o
o	$d,ct,i,m,$ e,cb,cv,o	i,cb,o	i,cb,o	$d,ct,m,$ cv,o
cv,o	ct,cv,o	$e,cb,cv,$ o	i,cb,o	ct,cv,o
ct,cv,o	ct,cv,o	ct,cv,o	$ct,i,e,$ cb,cv,o	ct,cv,o
o	$d,i,m,$ cb,o	i,cb,o	i,cb,o	$d,ct,i,m,$ e,cb,cv,o

表 2.11 $top(体,体) \wedge top(体,体)$

$\wedge$	$\begin{bmatrix} 0 & 0 & 1 \\ 0 & 0 & 1 \\ 1 & 1 & 1 \end{bmatrix}$ disjoint(B,C)	$\begin{bmatrix} 0 & 0 & 1 \\ 0 & 1 & 1 \\ 1 & 1 & 1 \end{bmatrix}$ meet(B,C)	$\begin{bmatrix} 1 & 0 & 0 \\ 0 & 1 & 0 \\ 0 & 0 & 1 \end{bmatrix}$ equal(B,C)
$\begin{bmatrix} 0 & 0 & 1 \\ 0 & 0 & 1 \\ 1 & 1 & 1 \end{bmatrix}$ disjoint	$d,m,e,$ $i,cb,ct,$ cv,o	$d,m,$ i,cb,o	d
$\begin{bmatrix} 0 & 0 & 1 \\ 0 & 1 & 1 \\ 1 & 1 & 1 \end{bmatrix}$ meet (A,B)	$d,m,ct,$ cv,o	$d,m,e,$ cb,cv,o	m
$\begin{bmatrix} 1 & 0 & 0 \\ 0 & 1 & 0 \\ 0 & 0 & 1 \end{bmatrix}$ equal (A,B)	d	m	e
$\begin{bmatrix} 1 & 0 & 0 \\ 1 & 0 & 0 \\ 1 & 1 & 1 \end{bmatrix}$ inside (A,B)	d	d	i
$\begin{bmatrix} 1 & 0 & 0 \\ 1 & 1 & 0 \\ 1 & 1 & 1 \end{bmatrix}$ coveredby (A,B)	d	d,m	cb
$\begin{bmatrix} 1 & 1 & 1 \\ 0 & 0 & 1 \\ 0 & 0 & 1 \end{bmatrix}$ contain (A,B)	$d,m,ct,$ cv,o	ct,cv,o	ct
$\begin{bmatrix} 1 & 1 & 1 \\ 0 & 1 & 1 \\ 0 & 0 & 1 \end{bmatrix}$ cover (A,B)	$d,m,ct,$ cv,o	$m,ct,$ cv,o	cv
$\begin{bmatrix} 1 & 1 & 1 \\ 1 & 1 & 1 \\ 1 & 1 & 1 \end{bmatrix}$ overlap (A,B)	$d,m,ct,$ cv,o	$d,m,ct,$ cv,o	o

表 2.12 $top(面,点) \wedge top(点,面)$

$\wedge$	$B \in C°$
$B \in A°$	PP3—PP8,PP11—PP14, PP17,PP18
$B \in \partial A$	PP3—PP5,PP11, PP15—PP18
$B \in A^-$	PP1—PP5,PP9—PP18

→*top*(体,体)

$\begin{bmatrix} 1 & 0 & 0 \\ 1 & 0 & 0 \\ 1 & 1 & 1 \end{bmatrix}$	$\begin{bmatrix} 1 & 0 & 0 \\ 1 & 1 & 0 \\ 1 & 1 & 1 \end{bmatrix}$	$\begin{bmatrix} 1 & 1 & 1 \\ 0 & 0 & 1 \\ 0 & 0 & 1 \end{bmatrix}$	$\begin{bmatrix} 1 & 1 & 1 \\ 0 & 1 & 1 \\ 0 & 0 & 1 \end{bmatrix}$	$\begin{bmatrix} 1 & 1 & 1 \\ 1 & 1 & 1 \\ 1 & 1 & 1 \end{bmatrix}$
inside(B,C)	coveredby(B,C)	contain(B,C)	cover(B,C)	overlap(B,C)
$d,m,$ i,cb,o	$d,m,i,$ cb,o	d	d	$d,m,i,$ cb,o
i,cb,o	$m,i,$ cb,o	d	d,m	$d,m,i,$ cb,o
i	cb	ct	cv	o
i	i	$d,m,e,$ $i,cb,ct,$ cv,o	$d,m,$ i,cb,o	$d,m,i,$ cb,o
i	i,cb	$d,m,ct,$ cv,o	$d,m,e,$ cb,cv,o	$d,m,i,$ cb,o
$e,i,cb,ct,$ cv,o	ct,cv,o	ct	ct	ct,cv,o
i,cb,o	$e,cb,$ cv,o	ct	ct,cv	ct,cv,o
i,cb,o	i,cb,o	$d,m,ct,$ cv,o	$d,m,ct,$ cv,o	$d,m,e,$ $i,cb,ct,$ cv,o

→*top*(面,面)

$B \in \partial C$	$B \in C^-$
PP3,PP6,PP7, PP9—PP13	PP1—PP3,PP6, PP7,PP9—PP18
PP2—PP4,PP6,PP8,PP9, PP12,PP14,PP15,PP17	PP1—PP3,PP6,PP7, PP9—PP18
PP1—PP5,PP9—PP18	PP1—PP18

表 2.13 $top(面,线) \wedge top(线,面)$

$\wedge$		LP1 $\begin{bmatrix} 0 & 0 & 1 \\ 0 & 0 & 1 \\ 1 & 1 & 1 \end{bmatrix}$	LP2 $\begin{bmatrix} 0 & 0 & 1 \\ 0 & 1 & 1 \\ 1 & 1 & 1 \end{bmatrix}$	LP3 $\begin{bmatrix} 0 & 1 & 1 \\ 0 & 1 & 1 \\ 1 & 1 & 1 \end{bmatrix}$	LP4 $\begin{bmatrix} 0 & 1 & 0 \\ 0 & 1 & 0 \\ 1 & 1 & 1 \end{bmatrix}$	LP5 $\begin{bmatrix} 0 & 1 & 1 \\ 0 & 0 & 1 \\ 1 & 1 & 1 \end{bmatrix}$	LP6 $\begin{bmatrix} 1 & 1 & 1 \\ 1 & 0 & 1 \\ 1 & 1 & 1 \end{bmatrix}$
$\begin{bmatrix} 0 & 0 & 1 \\ 0 & 0 & 1 \\ 1 & 1 & 1 \end{bmatrix}$	PL1	PP1—PP18	PP1—PP5, PP9—PP18	PP1—PP5, PP9—PP18	PP1—PP5, PP9—PP18	PP1—PP5, PP9—PP18	PP1—PP5, PP9—PP18
$\begin{bmatrix} 0 & 0 & 1 \\ 0 & 1 & 1 \\ 1 & 1 & 1 \end{bmatrix}$	PL2	PP1—PP3, PP6,PP7, PP9—PP18	PP1—PP4, PP6, PP8—PP18	PP1—PP5, PP9—PP18	PP2—PP4, PP9,PP12, PP14,PP15, PP17	PP1—PP3, PP9—PP18	PP1—PP5, PP9—PP18
$\begin{bmatrix} 0 & 0 & 1 \\ 1 & 1 & 1 \\ 1 & 1 & 1 \end{bmatrix}$	PL3	PP1—PP3, PP6,PP7, PP9—PP18	PP1—PP3, PP6, PP9—PP18	PP1—PP4, PP6, PP8—PP18	PP2—PP4, PP9,PP12, PP14,PP15, PP17	PP1—PP3, PP6, PP9—PP18	PP1—PP5, PP9—PP18
$\begin{bmatrix} 0 & 0 & 1 \\ 1 & 1 & 1 \\ 0 & 0 & 1 \end{bmatrix}$	PL4	PP1—PP3, PP6,PP7, PP9—PP18	PP2,PP3, PP6,PP9, PP12,PP14, PP15,PP17	PP2,PP3, PP6,PP9, PP12,PP14, PP15,PP17	PP2—PP4, PP6,PP8, PP9,PP12, PP14,PP15, PP17	PP2,PP3, PP6,PP9, PP12,PP14, PP15,PP17	PP3,PP15, PP17
$\begin{bmatrix} 0 & 0 & 1 \\ 1 & 0 & 1 \\ 1 & 1 & 1 \end{bmatrix}$	PL5	PP1—PP3, PP6,PP7, PP9—PP18	PP1—PP3, PP9—PP18	PP1—PP4, PP9—PP18	PP2—PP4, PP9,PP12, PP14,PP15, PP17	PP1—PP4, PP6, PP8—PP18	PP1—PP5, PP9—PP18
$\begin{bmatrix} 1 & 1 & 1 \\ 1 & 0 & 1 \\ 1 & 1 & 1 \end{bmatrix}$	PL6	PP1—PP3, PP6,PP7, PP9—PP18	PP1—PP3, PP6,PP7, PP9—PP18	PP1—PP3, PP6,PP7, PP9—PP18	PP3,PP9, PP12	PP1—PP3, PP6,PP7 PP9—PP18	PP1—PP18
$\begin{bmatrix} 1 & 1 & 1 \\ 0 & 1 & 1 \\ 0 & 0 & 1 \end{bmatrix}$	PL7	PP1—PP3, PP6,PP7, PP9—PP18	PP2,PP3, PP6,PP7, PP9—PP15, PP17	PP3,PP6, PP7, PP9—PP13	PP3,PP6, PP9,PP12	PP3,PP6, PP7, PP9—PP13	PP3,PP6, PP7, PP11—PP13
$\begin{bmatrix} 1 & 1 & 1 \\ 0 & 0 & 1 \\ 0 & 0 & 1 \end{bmatrix}$	PL8	PP1—PP3, PP6,PP7, PP9—PP18	PP3,PP6, PP7, PP9—PP13	PP3,PP6, PP7, PP9—PP13	PP3,PP6, PP7, PP9—PP13	PP3,PP6, PP7, PP9—PP13	PP3,PP6, PP7, PP11—PP13
$\begin{bmatrix} 1 & 0 & 1 \\ 0 & 1 & 1 \\ 0 & 0 & 1 \end{bmatrix}$	PL9	PP1—PP3, PP6,PP7, PP9—PP18	PP2,PP3, PP6,PP9, PP12,PP14, PP15,PP17	PP3,PP6, PP9,PP12	PP3,PP6, PP9,PP12	PP3,PP6, PP7, PP9—PP13	PP3, PP11
$\begin{bmatrix} 1 & 0 & 1 \\ 1 & 0 & 1 \\ 1 & 1 & 1 \end{bmatrix}$	PL10	PP1—PP3, PP6,PP7, PP9—PP18	PP1—PP3, PP9—PP18	PP1—PP3, PP9—PP18	PP3,PP9, PP12,	PP1—PP3, PP6,PP7, PP9—PP18	PP1—PP5, PP9—PP18

→*top*(面,面)

$\begin{bmatrix} 1 & 0 & 0 \\ 1 & 1 & 0 \\ 1 & 1 & 1 \end{bmatrix}$	$\begin{bmatrix} 1 & 0 & 0 \\ 1 & 0 & 0 \\ 1 & 1 & 1 \end{bmatrix}$	$\begin{bmatrix} 1 & 0 & 0 \\ 0 & 1 & 0 \\ 1 & 1 & 1 \end{bmatrix}$	$\begin{bmatrix} 1 & 1 & 1 \\ 0 & 0 & 1 \\ 1 & 1 & 1 \end{bmatrix}$	$\begin{bmatrix} 1 & 1 & 1 \\ 0 & 1 & 1 \\ 1 & 1 & 1 \end{bmatrix}$	$\begin{bmatrix} 0 & 0 & 1 \\ 1 & 0 & 1 \\ 1 & 1 & 1 \end{bmatrix}$	$\begin{bmatrix} 1 & 0 & 1 \\ 0 & 0 & 1 \\ 1 & 1 & 1 \end{bmatrix}$
LP7	LP8	LP9	LP10	LP11	LP12	LP13
PP1—PP5, PP9—PP18	PP1—PP5, PP9—PP18	PP1—PP5, PP9—PP18	PP1—PP5, PP9—PP18	PP1—PP5, PP9—PP18	PP1—PP5, PP9—PP18	PP1—PP5, PP9—PP18
PP2—PP5, PP9,PP11, PP12, PP14—PP18	PP3—PP5, PP11, PP15—PP18	PP2—PP4, PP9, PP12,PP14, PP15,PP17	PP1—PP3, PP9—PP18	PP1—PP4, PP9—PP18	PP1—PP5, PP9—PP18	PP1—PP3, PP9—PP18
PP3—PP5, PP11, PP15—PP18	PP3—PP5, PP11, PP15—PP18	PP3,PP4, PP15,PP17	PP1—PP3, PP9—PP18	PP1—PP4, PP9—PP18	PP1—PP3, PP9—PP18	PP1—PP3, PP9—PP18
PP3,PP4, PP15,PP17	PP3—PP5, PP11, PP15—PP18	PP3,PP4, PP15,PP17	PP3, PP15,PP17	PP3, PP15,PP17	PP3,PP11 PP15—PP18	PP3,PP11 PP15—PP18
PP3—PP5, PP11, PP15—PP18	PP3—PP5, PP11, PP15—PP18	PP3—PP5, PP11, PP15—PP18	PP1—PP5, PP9—PP18	PP1—PP5, PP9—PP18	PP1—PP3, PP9—PP18	PP1—PP5, PP9—PP18
PP3—PP5, PP11,PP17, PP18	PP3—PP5, PP11,PP17, PP18	PP3,PP11	PP1—PP3, PP6,PP7, PP9—PP18	PP1—PP3, PP6,PP7, PP9—PP18	PP1—PP3, PP6,PP7, PP9—PP18	PP1—PP3, PP6,PP7, PP9—PP18
PP3,PP4,PP6, PP8,PP11, PP12,PP14, PP17,PP18	PP3—PP5, PP11,PP17, PP18	PP3,PP6, PP12	PP3,PP6, PP7, PP11—PP13	PP3,PP6, PP7, PP11—PP13	PP3,PP6, PP7, PP11—PP18	PP3,PP6, PP7, PP11—PP14, PP17,PP18
PP3,PP6, PP7, PP11—PP13	PP3—PP8, PP11—PP14, PP17,PP18	PP3,PP6, PP7, PP11—PP13	PP3,PP6, PP7, PP11—PP13	PP3,PP6, PP7, PP11—PP13	PP3,PP6,PP7, PP11—PP14, PP17,PP18	PP3,PP6,PP7, PP11—PP14, PP17,PP18
PP3,PP4, PP17	PP3—PP5, PP11, PP17,PP18	PP3,PP4, PP6,PP8, PP12,PP14, PP17	PP3,PP6, PP7, PP11—PP13	PP3,PP6, PP12	PP3,PP11, PP15—PP18	PP3,PP6, PP7, PP11—PP14, PP17,PP18
PP3—PP5, PP11,PP17, PP18	PP3—PP5, PP11, PP17,PP18	PP3—PP5, PP11, PP17,PP18	PP1—PP13	PP1—PP5, PP9—PP18	PP1—PP3, PP9—PP18	PP1—PP7, PP9—PP18

∧	$\begin{bmatrix}0&0&1\\0&0&1\\1&1&1\end{bmatrix}$	$\begin{bmatrix}0&0&1\\0&1&1\\1&1&1\end{bmatrix}$	$\begin{bmatrix}0&1&1\\0&1&1\\1&1&1\end{bmatrix}$	$\begin{bmatrix}0&1&0\\0&1&0\\1&1&1\end{bmatrix}$	$\begin{bmatrix}0&1&1\\0&0&1\\1&1&1\end{bmatrix}$	$\begin{bmatrix}1&1&1\\1&0&1\\1&1&1\end{bmatrix}$
	LP1	LP2	LP3	LP4	LP5	LP6
$\begin{bmatrix}1&0&1\\1&1&1\\1&1&1\end{bmatrix}$ PL11	PP1—PP3, PP6,PP7, PP9—PP18	PP1—PP3, PP6, PP9—PP18	PP1—PP3, PP6, PP9—PP18	PP3,PP9, PP12,	PP1—PP3, PP6,PP7, PP9—PP18	PP1—PP5, PP9—PP18
$\begin{bmatrix}0&1&1\\0&0&1\\1&1&1\end{bmatrix}$ PL12	PP1—PP3, PP6,PP7, PP9—PP18	PP1—PP3, PP6,PP7, PP9—PP18	PP1—PP3, PP9—PP18	PP3, PP9—PP13	PP1—PP3, PP9—PP18	PP1—PP5, PP9—PP18
$\begin{bmatrix}1&0&1\\0&0&1\\1&1&1\end{bmatrix}$ PL13	PP1—PP3, PP6,PP7, PP9—PP18	PP1—PP3, PP9—PP18	PP1—PP3, PP9—PP18	PP3, PP9—PP13	PP1—PP3, PP6,PP7, PP9—PP18	PP1—PP5, PP9—PP18

表 2.14 *top*(面,面) ∧ *top*(面,面)

∧	$\begin{bmatrix}0&0&1\\0&0&1\\1&1&1\end{bmatrix}$	$\begin{bmatrix}0&0&1\\0&1&1\\1&1&1\end{bmatrix}$	$\begin{bmatrix}1&1&1\\1&1&1\\1&1&1\end{bmatrix}$	$\begin{bmatrix}1&0&0\\1&1&0\\1&1&1\end{bmatrix}$	$\begin{bmatrix}1&0&0\\1&0&0\\1&1&1\end{bmatrix}$	$\begin{bmatrix}1&1&1\\0&1&1\\0&0&1\end{bmatrix}$	$\begin{bmatrix}1&1&1\\0&0&1\\0&0&1\end{bmatrix}$	$\begin{bmatrix}1&0&0\\0&1&0\\0&0&1\end{bmatrix}$
	PP1	PP2	PP3	PP4	PP5	PP6	PP7	PP8
$\begin{bmatrix}0&0&1\\0&0&1\\1&1&1\end{bmatrix}$ PP1	PP1—PP18	PP1—PP5, PP9—PP18	PP1—PP5, PP9—PP18	PP1—PP5, PP9—PP18	PP1—PP5, PP9—PP18	PP1	PP1	PP1
$\begin{bmatrix}0&0&1\\0&1&1\\1&1&1\end{bmatrix}$ PP2	PP1—PP3, PP6,PP7, PP9—PP18	PP1—PP4, PP6, PP8—PP18	PP1—PP5, PP9—PP18	PP2—PP5, PP9,PP11, PP12, PP14—PP18	PP3—PP5, PP11, PP15—PP18	PP1,PP2	PP1	PP2
$\begin{bmatrix}1&1&1\\1&1&1\\1&1&1\end{bmatrix}$ PP3	PP1—PP3, PP6,PP7, PP9—PP18	PP1—PP3, PP6,PP7, PP9—PP18	PP1—PP18	PP3—PP5, PP11,PP17, PP18	PP3—PP5, PP11,PP17, PP18	PP1—PP3, PP6,PP7, PP9—PP18	PP1—PP3, PP6,PP7, PP9—PP18	PP3
$\begin{bmatrix}1&0&0\\1&1&0\\1&1&1\end{bmatrix}$ PP4	PP1	PP1,PP2	PP1—PP5, PP9—PP18	PP4,PP5	PP5	PP1—PP4, PP6, PP8—PP18	PP1—PP3, PP6,PP7, PP9—PP18	PP4
$\begin{bmatrix}1&0&0\\1&0&0\\1&1&1\end{bmatrix}$ PP5	PP1	PP1	PP1—PP5, PP9—PP18	PP5	PP5	PP1—PP5, PP9—PP18	PP1—PP18	PP5
$\begin{bmatrix}1&1&1\\0&1&1\\0&0&1\end{bmatrix}$ PP6	PP1—PP3, PP6,PP7, PP9—PP18	PP2,PP3, PP6,PP7, PP9—PP15, PP17	PP3,PP6, PP7, PP11—PP13	PP3,PP4,PP6, PP8,PP11, PP12,PP14, PP17,PP18	PP3—PP5, PP11,PP17, PP18	PP6,PP7	PP7	PP6

续表

$\begin{bmatrix}1&0&0\\1&1&0\\1&1&1\end{bmatrix}$	$\begin{bmatrix}1&0&0\\1&0&0\\1&1&1\end{bmatrix}$	$\begin{bmatrix}1&0&0\\0&1&0\\1&1&1\end{bmatrix}$	$\begin{bmatrix}1&1&1\\0&1&0\\1&1&1\end{bmatrix}$	$\begin{bmatrix}1&1&1\\0&0&1\\1&1&1\end{bmatrix}$	$\begin{bmatrix}0&0&1\\1&0&1\\1&1&1\end{bmatrix}$	$\begin{bmatrix}1&0&1\\0&0&1\\1&1&1\end{bmatrix}$
LP7	LP8	LP9	LP10	LP11	LP12	LP13
PP3—PP5, PP11,PP17, PP18	PP3—PP5, PP11,PP17, PP18	PP3,PP4, PP17	PP1—PP3, PP6,PP7, PP9—PP18	PP1—PP4, PP6, PP8—PP18	PP1—PP3, PP9—PP18	PP1—PP3, PP6,PP7, PP9—PP18
PP4,PP5, PP9—PP14, PP17,PP18	PP3—PP5, PP11—PP14, PP17,PP18	PP3, PP9—PP13	PP1—PP3, PP9—PP18	PP1—PP3, PP9—PP18	PP1—PP18	PP1—PP3, PP9—PP18
PP3—PP5, PP11—PP14, PP17,PP18	PP3—PP5, PP11,PP17, PP18	PP3—PP5, PP11—PP14, PP17,PP18	PP1—PP18	PP1—PP5, PP9—PP18	PP1—PP3, PP9—PP18	PP1—PP18

→ $top(面,面)$

$\begin{bmatrix}0&1&1\\0&1&1\\1&1&1\end{bmatrix}$	$\begin{bmatrix}0&1&1\\0&0&1\\1&1&1\end{bmatrix}$	$\begin{bmatrix}1&1&1\\1&0&1\\1&1&1\end{bmatrix}$	$\begin{bmatrix}1&1&1\\0&1&1\\1&1&1\end{bmatrix}$	$\begin{bmatrix}1&1&1\\0&0&1\\1&1&1\end{bmatrix}$	$\begin{bmatrix}1&0&1\\0&1&1\\1&1&1\end{bmatrix}$	$\begin{bmatrix}0&0&1\\1&1&1\\1&1&1\end{bmatrix}$	$\begin{bmatrix}0&0&1\\1&0&1\\1&1&1\end{bmatrix}$	$\begin{bmatrix}1&0&1\\1&1&1\\1&1&1\end{bmatrix}$	$\begin{bmatrix}1&0&1\\1&0&1\\1&1&1\end{bmatrix}$
PP9	PP10	PP11	PP12	PP13	PP14	PP15	PP16	PP17	PP18
PP1—PP5, PP9—PP18	PP1—PP5, PP9—PP18	PP1—PP5, PP9—PP18	PP1—PP5, PP9—PP18	PP1—PP5, PP9—PP18	PP1—PP5, PP9—PP18	PP1—PP5, PP9—PP18	PP1—PP5, PP9—PP18	PP1—PP5, PP9—PP18	PP1—PP5, PP9—PP18
PP1—PP5, PP9—PP18	PP1—PP3, PP9—PP18	PP1—PP5, PP9—PP18	PP1—PP4, PP9—PP18	PP1—PP3, PP9—PP18	PP1—PP4, PP9—PP18	PP1—PP5, PP9—PP18	PP1—PP5, PP9—PP18	PP1—PP5, PP9—PP18	PP1—PP5, PP9—PP18
PP1—PP3, PP6,PP7, PP9—PP18	PP1—PP3, PP6,PP7, PP9—PP18	PP1—PP7, PP9—PP18	PP1—PP3, PP6,PP7, PP9—PP18	PP1—PP3, PP6,PP7, PP9—PP18	PP1—PP3, PP6,PP7, PP9—PP18	PP1—PP3, PP6,PP7, PP9—PP18	PP1—PP3, PP6,PP7, PP9—PP18	PP1—PP7, PP9—PP18	PP1—PP7, PP9—PP18
PP1,PP2, PP9,PP10	PP1,PP2, PP9,PP10	PP1—PP5, PP9—PP18	PP1—PP4, PP9—PP18	PP1—PP3, PP9—PP18	PP1,PP2, PP4, PP14—PP18	PP1,PP2, PP15,PP16	PP1,PP16	PP1,PP2, PP4,PP5, PP14—PP18	PP1,PP5, PP16,PP18
PP1,PP2, PP9,PP10	PP1,PP2, PP9,PP10	PP1—PP5, PP9—PP18	PP1—PP5, PP9—PP18	PP1—PP5, PP9—PP18	PP1,PP5, PP16,PP18	PP1	PP1	PP1,PP5, PP16,PP18	PP1,PP5, PP16,PP18
PP3,PP6, PP7, PP9—PP13	PP3,PP6, PP7, PP9—PP13	PP3,PP6, PP7, PP11—PP13	PP3,PP6, PP7, PP11—PP13	PP3,PP6, PP7, PP11—PP13	PP3,PP6, PP7, PP11—PP14, PP17,PP18	PP3,PP6, PP7, PP11—PP15, PP17	PP3,PP6, PP7, PP11—PP18	PP3,PP6, PP7, PP11—PP14, PP17	PP3,PP6, PP7, PP11—PP14, PP17,PP18

∧	$\begin{bmatrix}0&0&1\\0&0&1\\1&1&1\end{bmatrix}$ PP1	$\begin{bmatrix}0&0&1\\0&1&1\\1&1&1\end{bmatrix}$ PP2	$\begin{bmatrix}1&1&1\\1&1&1\\1&1&1\end{bmatrix}$ PP3	$\begin{bmatrix}1&0&0\\1&1&0\\1&1&1\end{bmatrix}$ PP4	$\begin{bmatrix}1&0&0\\1&0&0\\1&1&1\end{bmatrix}$ PP5	$\begin{bmatrix}1&1&1\\0&1&1\\0&0&1\end{bmatrix}$ PP6	$\begin{bmatrix}1&1&1\\0&0&1\\0&0&1\end{bmatrix}$ PP7	$\begin{bmatrix}1&0&0\\0&1&0\\0&0&1\end{bmatrix}$ PP8
$\begin{bmatrix}1&1&1\\0&0&1\\0&0&1\end{bmatrix}$ PP7	PP1—PP3, PP6,PP7, PP9—PP18	PP3,PP6, PP7, PP9—PP13	PP3,PP6, PP7, PP11—PP13	PP3,PP6, PP7, PP11—PP13	PP3—PP8, PP11—PP14, PP17,PP18	PP7	PP7	PP7
$\begin{bmatrix}1&0&0\\0&1&0\\0&0&1\end{bmatrix}$ PP8	PP1	PP2	PP3	PP4	PP5	PP6	PP7	PP8
$\begin{bmatrix}0&1&1\\0&1&1\\1&1&1\end{bmatrix}$ PP9	PP1—PP3, PP6,PP7, PP9—PP18	PP1—PP3, PP6,PP7, PP9—PP18	PP1—PP5, PP9—PP18	PP3—PP5, PP9,PP11, PP12,PP14, PP17,PP18	PP3—PP5, PP11,PP17, PP18	PP1,PP2, PP9,PP10	PP1	PP9
$\begin{bmatrix}0&1&1\\0&0&1\\1&1&1\end{bmatrix}$ PP10	PP1—PP3, PP6,PP7, PP9—PP18	PP1—PP3, PP6,PP7, PP9—PP18	PP1—PP5, PP9—PP18	PP1—PP5, PP9—PP14, PP17,PP18	PP3—PP5, PP11—PP14, PP17,PP18	PP1,PP10	PP1	PP10
$\begin{bmatrix}1&1&1\\1&0&1\\1&1&1\end{bmatrix}$ PP11	PP1—PP3, PP6,PP7, PP9—PP18	PP1—PP3, PP6,PP7, PP9—PP18	PP1—PP7, PP9—PP18	PP3—PP5, PP11,PP17, PP18	PP3—PP5, PP11,PP17, PP18	PP1—PP3, PP6,PP7, PP9—PP18	PP1—PP3, PP6,PP7, PP9—PP18	PP11
$\begin{bmatrix}1&1&1\\0&0&1\\1&1&1\end{bmatrix}$ PP12	PP1—PP3, PP6,PP7, PP9—PP18	PP1—PP3, PP6,PP7, PP9—PP18	PP1—PP7, PP9—PP18	PP3—PP5, PP11,PP12, PP14,PP17, PP18	PP3—PP5, PP11,PP17, PP18	PP1,PP2, PP6,PP7, PP9,PP10, PP12—PP14	PP1,PP7, PP10,PP13	PP12
$\begin{bmatrix}1&1&1\\0&0&1\\1&1&1\end{bmatrix}$ PP13	PP1—PP3, PP6,PP7, PP9—PP18	PP1—PP3, PP6,PP7, PP9—PP18	PP1—PP7, PP9—PP18	PP3—PP5, PP11—PP14, PP17,PP18	PP3—PP5, PP11—PP14, PP17,PP18	PP1,PP7, PP10,PP13	PP1,PP7, PP10,PP13	PP13
$\begin{bmatrix}1&0&1\\0&1&1\\1&1&1\end{bmatrix}$ PP14	PP1—PP3, PP6,PP7, PP9—PP18	PP1—PP3, PP6, PP9—PP18	PP1—PP5, PP9—PP18	PP3—PP5, PP11,PP12, PP14,PP17, PP18	PP3—PP5, PP11,PP17, PP18	PP1—PP3, PP6, PP9—PP18	PP1,PP7, PP10,PP13	PP14
$\begin{bmatrix}0&0&1\\1&1&1\\1&1&1\end{bmatrix}$ PP15	PP1—PP3, PP6,PP7, PP9—PP18	PP1—PP3, PP6, PP9—PP18	PP1—PP5, PP9—PP18	PP3—PP5, PP11, PP15—PP18	PP3—PP5, PP11, PP15—PP18	PP1,PP2, PP15,PP16	PP1—PP3, PP9—PP18	PP15
$\begin{bmatrix}0&0&1\\1&1&1\\1&1&1\end{bmatrix}$ PP16	PP1—PP3, PP6,PP7, PP9—PP18	PP1—PP3, PP9—PP18	PP1—PP5, PP9—PP18	PP3—PP5, PP11, PP15—PP18	PP3—PP5, PP11, PP15—PP18	PP1,PP2, PP15,PP16	PP1,PP2, PP15,PP16	PP16
$\begin{bmatrix}1&0&1\\1&1&1\\1&1&1\end{bmatrix}$ PP17	PP1—PP3, PP6,PP7, PP9—PP18	PP1—PP3, PP6, PP9—PP18	PP1—PP5 PP9—PP18	PP3—PP5, PP11,PP17, PP18	PP3—PP5, PP11,PP17, PP18	PP1—PP3, PP6, PP9—PP18	PP1—PP3, PP6,PP7, PP9—PP18	PP17
$\begin{bmatrix}1&0&1\\1&0&1\\1&1&1\end{bmatrix}$ PP18	PP1—PP3, PP6,PP7, PP9—PP18	PP1—PP3, PP9—PP18	PP1—PP5, PP9—PP18	PP3—PP5, PP11,PP17, PP18	PP3—PP5, PP11,PP17, PP18	PP1—PP3, PP9—PP18	PP1—PP3, PP6,PP7, PP9—PP18	PP18

第 2 章 拓扑关系描述及其定性推理 47

续表

$\begin{bmatrix}0&1&1\\0&1&1\\1&1&1\end{bmatrix}$	$\begin{bmatrix}0&1&1\\0&0&1\\1&1&1\end{bmatrix}$	$\begin{bmatrix}1&1&1\\1&0&1\\1&1&1\end{bmatrix}$	$\begin{bmatrix}1&1&1\\0&1&1\\1&1&1\end{bmatrix}$	$\begin{bmatrix}1&1&1\\0&0&1\\1&1&1\end{bmatrix}$	$\begin{bmatrix}1&0&1\\0&1&1\\1&1&1\end{bmatrix}$	$\begin{bmatrix}0&0&1\\1&1&1\\1&1&1\end{bmatrix}$	$\begin{bmatrix}0&0&1\\1&0&1\\1&1&1\end{bmatrix}$	$\begin{bmatrix}1&0&1\\1&1&1\\1&1&1\end{bmatrix}$	$\begin{bmatrix}1&0&1\\1&0&1\\1&1&1\end{bmatrix}$
PP9	PP10	PP11	PP12	PP13	PP14	PP15	PP16	PP17	PP18
PP3,PP6, PP7, PP9—PP13	PP3,PP6, PP7, PP9—PP13	PP3,PP6, PP7, PP11—PP13	PP3,PP6, PP7, PP11—PP13	PP3,PP6, PP7, PP11—PP13	PP3,PP6, PP7, PP11—PP13	PP3,PP6, PP7, PP11—PP13	PP3,PP6,PP7, PP11—PP14, PP17,PP18	PP3,PP6, PP7, PP11—PP13	PP3,PP6,PP7, PP11—PP14, PP17,PP18
PP9	PP10	PP11	PP12	PP13	PP14	PP15	PP16	PP17	PP18
PP1—PP3, PP9—PP18	PP1—PP3, PP9—PP18	PP1—PP5, PP9—PP18	PP1—PP3, PP9—PP18	PP1—PP3, PP9—PP18	PP1—PP3, PP9—PP18	PP1—PP18	PP1—PP7, PP9—PP18	PP1—PP5, PP9—PP18	PP1—PP5, PP9—PP18
PP1—PP3, PP9—PP18	PP1—PP3, PP9—PP18	PP1—PP5, PP9—PP18	PP1—PP3, PP9—PP18	PP1—PP3, PP9—PP18	PP1—PP5, PP9—PP18	PP1—PP7, PP9—PP18	PP1—PP18	PP1—PP5, PP9—PP18	PP1—PP5, PP9—PP18
PP1—PP3, PP6,PP7, PP9—PP18	PP1—PP3, PP6,PP7, PP9—PP18	PP1—PP18	PP1—PP3, PP6,PP7, PP9—PP18	PP1—PP3, PP6,PP7, PP9—PP18	PP1—PP3, PP6,PP7, PP9—PP18	PP1—PP3, PP6,PP7, PP9—PP18	PP1—PP3, PP6,PP7, PP9—PP18	PP1—PP7, PP9—PP18	PP1—PP7, PP9—PP18
PP1—PP3, PP6,PP7, PP9—PP18	PP1—PP3, PP6,PP7, PP9—PP18	PP1—PP7, PP9—PP18	PP1—PP3, PP6,PP7, PP9—PP18	PP1—PP3, PP6,PP7, PP9—PP18	PP1—PP3, PP6,PP7, PP9—PP18	PP1—PP3, PP6,PP7, PP9—PP18	PP1—PP3, PP6,PP7, PP9—PP18	PP1—PP7, PP9—PP18	PP1—PP7, PP9—PP18
PP1—PP3, PP6,PP7, PP9—PP18	PP1—PP3, PP6,PP7, PP9—PP18	PP1—PP7, PP9—PP18	PP1—PP3, PP6,PP7, PP9—PP18	PP1—PP3, PP6,PP7, PP9—PP18	PP1—PP3, PP6,PP7, PP9—PP18	PP1—PP3, PP6,PP7, PP9—PP18	PP1—PP3, PP6,PP7, PP9—PP18	PP1—PP7, PP9—PP18	PP1—PP18
PP1—PP3, PP6, PP9—PP18	PP1—PP3, PP6,PP7, PP9—PP18	PP1—PP5, PP9—PP18	PP1—PP4, PP6, PP9—PP18	PP1—PP3, PP6,PP7, PP9—PP18	PP1—PP4, PP6, PP8—PP18	PP1—PP3, PP9—PP18	PP1—PP3, PP9—PP18	PP1—PP5, PP9—PP18	PP1—PP5, PP9—PP18
PP1—PP4, PP6, PP8—PP18	PP1—PP3, PP6, PP9—PP18	PP1—PP5, PP9—PP18	PP1—PP4, PP9—PP18	PP1—PP3, PP9—PP18	PP1—PP4, PP9—PP18	PP1—PP3, PP9—PP18	PP1—PP3, PP9—PP18	PP1—PP5, PP9—PP18	PP1—PP5, PP9—PP18
PP1—PP4, PP9—PP18	PP1—PP4, PP6, PP8—PP18	PP1—PP5, PP9—PP18	PP1—PP5, PP9—PP18	PP1—PP5, PP9—PP18	PP1—PP5, PP9—PP18	PP1—PP3, PP9—PP18	PP1—PP3, PP9—PP18	PP1—PP5, PP9—PP18	PP1—PP5, PP9—PP18
PP1—PP3, PP6, PP9—PP18	PP1—PP3, PP6,PP7, PP9—PP18	PP1—PP5, PP9—PP18	PP1—PP4, PP6, PP8—PP18	PP1—PP3, PP6,PP7, PP9—PP18	PP1—PP4, PP6, PP9—PP18	PP1—PP3, PP9—PP18	PP1—PP3, PP9—PP18	PP1—PP5, PP9—PP18	PP1—PP5, PP9—PP18
PP1—PP3, PP9—PP18	PP1—PP3, PP6,PP7, PP9—PP18	PP1—PP5, PP9—PP18	PP1—PP5, PP9—PP18	PP1—PP18	PP1—PP5, PP9—PP18	PP1—PP3, PP9—PP18	PP1—PP3, PP9—PP18	PP1—PP5, PP9—PP18	PP1—PP5, PP9—PP18

表 2.15　$top(面,体) \wedge top(体,面)$

$\wedge$	$\begin{bmatrix} 0 & 0 & 1 \\ 0 & 0 & 1 \\ 1 & 1 & 1 \end{bmatrix}$ TP1	$\begin{bmatrix} 0 & 0 & 1 \\ 0 & 1 & 1 \\ 1 & 1 & 1 \end{bmatrix}$ TP2	$\begin{bmatrix} 0 & 0 & 1 \\ 1 & 0 & 1 \\ 1 & 1 & 1 \end{bmatrix}$ TP3	$\begin{bmatrix} 0 & 0 & 1 \\ 1 & 1 & 1 \\ 1 & 1 & 1 \end{bmatrix}$ TP4
$\begin{bmatrix} 0 & 0 & 1 \\ 0 & 0 & 1 \\ 1 & 1 & 1 \end{bmatrix}$ PT1	PP1—PP18	PP1—PP5, PP9—PP18	PP1—PP5, PP9—PP18	PP1—PP5, PP9—PP18
$\begin{bmatrix} 0 & 0 & 1 \\ 0 & 1 & 1 \\ 1 & 1 & 1 \end{bmatrix}$ PT2	PP1—PP3, PP6,PP7, PP9—PP18	PP1—PP4, PP6, PP8—PP18	PP1—PP5, PP9—PP18	PP1—PP5, PP9—PP18
$\begin{bmatrix} 0 & 1 & 1 \\ 0 & 0 & 1 \\ 1 & 1 & 1 \end{bmatrix}$ PT3	PP1—PP3, PP6,PP7, PP9—PP18	PP1—PP3, PP6,PP7, PP9—PP18	PP1—PP5, PP6—PP8, PP9—PP18	PP1—PP7, PP9—PP18
$\begin{bmatrix} 0 & 1 & 1 \\ 0 & 1 & 1 \\ 1 & 1 & 1 \end{bmatrix}$ PT4	PP1—PP3, PP9—PP18	PP1—PP3, PP6,PP7, PP9—PP18	PP1—PP7, PP9—PP18	PP1—PP18
$\begin{bmatrix} 0 & 1 & 0 \\ 0 & 1 & 0 \\ 1 & 1 & 1 \end{bmatrix}$ PT5	PP1	PP1,PP2, PP9,PP10	PP1,PP5, PP16,PP18	PP1—PP5, PP9—PP18
$\begin{bmatrix} 1 & 1 & 1 \\ 1 & 1 & 1 \\ 1 & 1 & 1 \end{bmatrix}$ PT6	PP1—PP3, PP6,PP7, PP9—PP18	PP1—PP3, PP6,PP7, PP9—PP18	PP1—PP3, PP6,PP7, PP9—PP18	PP1—PP3, PP6,PP7, PP9—PP18
$\begin{bmatrix} 1 & 0 & 0 \\ 1 & 1 & 0 \\ 1 & 1 & 1 \end{bmatrix}$ PT7	PP1	PP1,PP2	PP1,PP16	PP1,PP2, PP15,PP16
$\begin{bmatrix} 1 & 0 & 0 \\ 1 & 0 & 0 \\ 1 & 1 & 1 \end{bmatrix}$ PT8	PP1	PP1	PP1	PP1
$\begin{bmatrix} 1 & 1 & 1 \\ 0 & 0 & 1 \\ 1 & 1 & 1 \end{bmatrix}$ PT9	PP1—PP3, PP6,PP7, PP9—PP18	PP1—PP3, PP6,PP7, PP9—PP18	PP1—PP3, PP6,PP7, PP9—PP18	PP1—PP3, PP6,PP7, PP9—PP18

表 2.16　$top(线,点) \wedge top(点,线)$

$\wedge$	$B \in C^\circ$
$B \in A^\circ$	LL3—LL9
$B \in \partial A$	LL3—LL5,LL10
$B \in A^-$	LL1—LL5,LL9—LL11

→ $top(面,面)$

$\begin{bmatrix} 0 & 0 & 1 \\ 1 & 1 & 1 \\ 0 & 0 & 1 \end{bmatrix}$	$\begin{bmatrix} 1 & 1 & 1 \\ 1 & 1 & 1 \\ 1 & 1 & 1 \end{bmatrix}$	$\begin{bmatrix} 1 & 1 & 1 \\ 0 & 1 & 1 \\ 0 & 0 & 1 \end{bmatrix}$	$\begin{bmatrix} 1 & 1 & 1 \\ 0 & 0 & 1 \\ 0 & 0 & 1 \end{bmatrix}$	$\begin{bmatrix} 1 & 0 & 1 \\ 1 & 0 & 1 \\ 1 & 1 & 1 \end{bmatrix}$
TP5	TP6	TP7	TP8	TP9
PP1	PP1—PP5, PP9—PP18	PP1	PP1	PP1—PP5, PP9—PP18
PP1,PP2, PP15,PP16	PP1—PP5, PP9—PP18	PP1,PP2	PP1	PP1—PP5, PP9—PP18
PP1,PP7, PP10,PP13	PP1—PP5, PP9—PP18	PP1,PP10	PP1	PP1—PP5, PP9—PP18
PP1—PP3, PP6,PP7, PP9—PP18	PP1—PP5, PP9—PP18	PP1,PP2, PP9,PP10	PP1	PP1—PP5, PP9—PP18
PP1—PP18	PP1—PP5, PP9—PP18	PP1—PP3, PP9—PP18	PP1	PP1,PP5, PP16,PP18
PP1—PP3, PP6,PP7, PP9—PP18	PP1—PP18	PP1—PP3, PP6,PP7, PP9—PP18	PP1—PP3, PP6,PP7, PP9—PP18	PP1,PP7, PP9—PP18
PP1,PP2, PP15,PP16	PP1—PP5, PP9—PP18	PP1—PP4,PP6, PP8—PP18	PP1—PP3, PP6,PP7, PP9—PP18	PP1,PP5, PP16,PP18
PP1	PP1—PP5, PP9—PP18	PP1—PP5, PP9—PP18	PP1—PP18	PP1,PP5, PP16—PP18
PP1,PP7, PP10,PP13	PP1—PP7, PP9—PP18	PP1,PP7, PP10,PP13	PP1,PP7, PP10,PP13	PP1—PP18

→ $top(线,线)$

$B \in \partial C$	$B \in C^-$
LL3,LL7,LL8,LL11	LL1—LL3,LL7—LL11
LL2,LL4,LL6,LL7	LL1—LL3,LL7—LL11
LL1—LL5,LL9—LL11	LL1—LL11

表 2.17　$top(线,线) \wedge top(线,线)$

$\wedge$		$\begin{bmatrix}0&0&1\\0&0&1\\1&1&1\end{bmatrix}$	$\begin{bmatrix}0&0&1\\0&1&1\\1&1&1\end{bmatrix}$	$\begin{bmatrix}1&1&1\\1&0&1\\1&1&1\end{bmatrix}$	$\begin{bmatrix}1&0&0\\1&1&0\\1&1&1\end{bmatrix}$	$\begin{bmatrix}1&0&0\\1&0&0\\1&1&1\end{bmatrix}$
		LL1	LL2	LL3	LL4	LL5
$\begin{bmatrix}0&0&1\\0&0&1\\1&1&1\end{bmatrix}$	LL1	LL1—LL11	LL1—LL5, LL9—LL11	LL1—LL5, LL9—LL11	LL1—LL5, LL9—LL11	LL1—LL5, LL9—LL11
$\begin{bmatrix}0&0&1\\0&1&1\\1&1&1\end{bmatrix}$	LL2	LL1—LL3, LL7—LL11	LL1—LL4, LL6, LL7, LL9—LL11	LL1—LL5, LL9—LL11	LL2—LL5, LL10	LL3—LL5, LL10
$\begin{bmatrix}1&1&1\\1&0&1\\1&1&1\end{bmatrix}$	LL3	LL1—LL3, LL7—LL11	LL1—LL3, LL7—LL11	LL1—LL11	LL3—LL5	LL3—LL5
$\begin{bmatrix}1&0&0\\1&1&0\\1&1&1\end{bmatrix}$	LL4	LL1	LL1, LL2	LL1—LL5, LL9—LL11	LL4, LL5	LL5
$\begin{bmatrix}0&0&0\\1&0&0\\1&1&1\end{bmatrix}$	LL5	LL1	LL1	LL1—LL5, LL9—LL11	LL5	LL5
$\begin{bmatrix}1&0&0\\0&1&0\\0&0&1\end{bmatrix}$	LL6	LL1	LL2	LL3	LL4	LL5
$\begin{bmatrix}1&1&1\\0&1&1\\0&0&1\end{bmatrix}$	LL7	LL1—LL3, LL7—LL11	LL2, LL3, LL7, LL8, LL11	LL3, LL7, LL8	LL3, LL4, LL6, LL7	LL3—LL5
$\begin{bmatrix}1&1&1\\0&0&1\\0&0&1\end{bmatrix}$	LL8	LL1—LL3, LL7—LL11	LL3, LL7, LL8, LL11	LL3, LL7, LL8	LL3, LL7, LL8	LL3—LL9
$\begin{bmatrix}1&0&1\\0&0&1\\1&1&1\end{bmatrix}$	LL9	LL1—LL3, LL7—LL11	LL1—LL3, LL9—LL11	LL1—LL5, LL9—LL11	LL3—LL5, LL9	LL3—LL5, LL9
$\begin{bmatrix}0&0&1\\1&0&1\\1&1&1\end{bmatrix}$	LL10	LL1—LL3, LL7—LL11	LL1—LL3, LL9—LL11	LL1—LL5, LL9—LL11	LL3—LL5, LL10	LL3—LL5, LL10
$\begin{bmatrix}0&1&1\\0&0&1\\1&1&1\end{bmatrix}$	LL11	LL1—LL3, LL7—LL11	LL1—LL3, LL7—LL11	LL1—LL5, LL9—LL11	LL3—LL5, LL9, LL11	LL3—LL5, LL9

→top(线,线)

$\begin{bmatrix} 1 & 0 & 0 \\ 0 & 1 & 0 \\ 0 & 0 & 1 \end{bmatrix}$	$\begin{bmatrix} 1 & 1 & 1 \\ 0 & 1 & 1 \\ 0 & 0 & 1 \end{bmatrix}$	$\begin{bmatrix} 1 & 1 & 1 \\ 0 & 0 & 1 \\ 0 & 0 & 1 \end{bmatrix}$	$\begin{bmatrix} 1 & 0 & 1 \\ 0 & 0 & 1 \\ 1 & 1 & 1 \end{bmatrix}$	$\begin{bmatrix} 0 & 0 & 1 \\ 1 & 0 & 1 \\ 1 & 1 & 1 \end{bmatrix}$	$\begin{bmatrix} 0 & 1 & 1 \\ 0 & 0 & 1 \\ 1 & 1 & 1 \end{bmatrix}$
LL6	LL7	LL8	LL9	LL10	LL11
LL1	LL1	LL1	LL1—LL5, LL9—LL11	LL1—LL5, LL9—LL11	LL1—LL5, LL9—LL11
LL2	LL1,LL2	LL1	LL1—LL3, LL9—LL11	LL1—LL5, LL9—LL11	LL1—LL3, LL9—LL11
LL3	LL1—LL3, LL7—LL11	LL1—LL3, LL7—LL11	LL1—LL3, LL7—LL11	LL1—LL3, LL7—LL11	LL1—LL3, LL7—LL11
LL4	LL1—LL4, LL6,LL7 LL9—LL11	LL1—LL3, LL7—LL11	LL1,LL9, LL10	LL1,LL10	LL1,LL2, LL11
LL5	LL1—LL5, LL9—LL11	LL1—LL11	LL1,LL5, LL9,LL10	LL1	LL1,LL2, LL11
LL6	LL7	LL8	LL9	LL10	LL11
LL7	LL7,LL8	LL8	LL3, LL7—LL9	LL3, LL7—LL10	LL3,LL7, LL8,LL11
LL8	LL8	LL8	LL3, LL7—LL9	LL3, LL7—LL9	LL3,LL7, LL8,LL11
LL9	LL1,LL9, LL11	LL1,LL8, LL9,LL11	LL1—LL11	LL1—LL3, LL9—LL11	LL1—LL3, LL7—LL11
LL10	LL1,LL2, LL10	LL1,LL2, LL10	LL1—LL5, LL9—LL11	LL1—LL3, LL9—LL11	LL1—LL4, LL6,LL7 LL9—LL11
LL11	LL1,LL11	LL1	LL1—LL3, LL9—LL11	LL1—LL11	LL1—LL3, LL9—LL11

表 2.18 $top(线,面) \wedge top(面,线)$

$\wedge$	$\begin{bmatrix} 0&0&1\\0&0&1\\1&1&1 \end{bmatrix}$ PL1	$\begin{bmatrix} 0&0&1\\0&1&1\\1&1&1 \end{bmatrix}$ PL2	$\begin{bmatrix} 0&0&1\\1&1&1\\1&1&1 \end{bmatrix}$ PL3	$\begin{bmatrix} 0&0&1\\1&0&1\\0&0&1 \end{bmatrix}$ PL4	$\begin{bmatrix} 0&0&1\\1&0&1\\1&1&1 \end{bmatrix}$ PL5	$\begin{bmatrix} 1&1&1\\1&0&1\\1&1&1 \end{bmatrix}$ PL6
$\begin{bmatrix} 0&0&1\\0&0&1\\1&1&1 \end{bmatrix}$ LP1	LL1—LL11	LL1—LL5, LL9—LL11	LL1—LL5, LL9—LL11	LL1	LL1—LL5, LL9—LL11	LL1—LL5, LL9—LL11
$\begin{bmatrix} 0&0&1\\0&1&1\\1&1&1 \end{bmatrix}$ LP2	LL1—LL3, LL7—LL11	LL1—LL4, LL6,LL7, LL9—LL11	LL1—LL5, LL9—LL11	LL1,LL2, LL10	LL1—LL5, LL9—LL11	LL1—LL5, LL9—LL11
$\begin{bmatrix} 0&1&1\\0&1&1\\1&1&1 \end{bmatrix}$ LP3	LL1—LL3, LL7—LL11	LL1—LL3, LL7—LL11	LL1—LL11	LL1—LL3, LL7—LL11	LL1—LL5, LL9—LL11	LL1—LL5, LL9—LL11
$\begin{bmatrix} 0&1&0\\0&1&0\\1&1&1 \end{bmatrix}$ LP4	LL1	LL1,LL2, LL11	LL1—LL5, LL9—LL11	LL1—LL11	LL1,LL5, LL9,LL10	LL1,LL9, LL10
$\begin{bmatrix} 0&1&1\\0&0&1\\1&1&1 \end{bmatrix}$ LP5	LL1—LL3, LL7—LL11	LL1—LL3, LL7—LL11	LL1—LL5, LL7—LL11	LL1,LL8, LL9,LL11	LL1—LL11	LL1—LL5, LL9—LL11
$\begin{bmatrix} 1&1&1\\1&0&1\\1&1&1 \end{bmatrix}$ LP6	LL1—LL3, LL7—LL11	LL1—LL3, LL7—LL11	LL1—LL3, LL7—LL11	LL1,LL8, LL9,LL11	LL1—LL3, LL7—LL11	LL1—LL11
$\begin{bmatrix} 1&0&0\\1&1&0\\1&1&1 \end{bmatrix}$ LP7	LL1	LL1,LL2	LL1,LL2, LL10	LL1,LL2, LL10	LL1,LL10	LL1—LL5, LL9—LL11
$\begin{bmatrix} 1&0&0\\1&0&0\\1&1&1 \end{bmatrix}$ LP8	LL1	LL1	LL1	LL1	LL1	LL1—LL5, LL9—LL11
$\begin{bmatrix} 1&0&0\\0&1&0\\1&1&1 \end{bmatrix}$ LP9	LL1	LL1,LL2	LL1,LL2, LL10	LL1,LL2, LL10	LL1,LL10	LL1,LL3, LL5, LL9—LL11
$\begin{bmatrix} 1&1&1\\0&0&1\\1&1&1 \end{bmatrix}$ LP10	LL1—LL3, LL7—LL11	LL1—LL3, LL7—LL11	LL1—LL3, LL7—LL11	LL1,LL8, LL9,LL11	LL1—LL3, LL7—LL11	LL1—LL5, LL7—LL11
$\begin{bmatrix} 1&1&1\\0&1&1\\1&1&1 \end{bmatrix}$ LP11	LL1—LL3, LL7—LL11	LL1—LL3, LL7—LL11	LL1—LL3, LL7—LL11	LL1—LL3, LL7—LL11	LL1—LL3, LL7—LL11	LL1—LL5, LL7—LL11
$\begin{bmatrix} 0&0&1\\1&0&1\\1&1&1 \end{bmatrix}$ LP12	LL1—LL3, LL7—LL11	LL1—LL3, LL9—LL11	LL1—LL3, LL9—LL11	LL1	LL1—LL3, LL9—LL11	LL1—LL5, LL9—LL11
$\begin{bmatrix} 1&0&1\\0&0&1\\1&1&1 \end{bmatrix}$ LP13	LL1—LL3, LL7—LL11	LL1—LL3, LL9—LL11	LL1—LL3, LL9—LL11	LL1	LL1—LL3, LL9—LL11	LL1—LL5, LL9—LL11

→*top*(线,线)

$\begin{bmatrix}1&1&1\\0&1&1\\0&0&1\end{bmatrix}$	$\begin{bmatrix}1&1&1\\0&0&1\\0&0&1\end{bmatrix}$	$\begin{bmatrix}1&0&1\\0&1&1\\0&0&1\end{bmatrix}$	$\begin{bmatrix}1&0&1\\1&0&1\\1&1&1\end{bmatrix}$	$\begin{bmatrix}1&0&1\\1&1&1\\1&1&1\end{bmatrix}$	$\begin{bmatrix}0&1&1\\0&0&1\\1&1&1\end{bmatrix}$	$\begin{bmatrix}1&0&1\\0&0&1\\1&1&1\end{bmatrix}$
PL7	PL8	PL9	PL10	PL11	PL12	PL13
LL1	LL1	LL1	LL1—LL5, LL9—LL11	LL1—LL5, LL9—LL11	LL1—LL5, LL9—LL11	LL1—LL5, LL9—LL11
LL1,LL2	LL1	LL1,LL2	LL1—LL5, LL9—LL11	LL1—LL5, LL9—LL11	LL1—LL3, LL9—LL11	LL1—LL3, LL9—LL11
LL1,LL2, LL11	LL1	LL1,LL2, LL11	LL1—LL5, LL9—LL11	LL1—LL5, LL9—LL11	LL1—LL3, LL9—LL11	LL1—LL3, LL9—LL11
LL1,LL2, LL11	LL1	LL1,LL2, LL11	LL1,LL5 LL9,LL10	LL1—LL5, LL9—LL11	LL1	LL1
LL1,LL11	LL1	LL1,LL11	LL1—LL5, LL9—LL11	LL1—LL5, LL9—LL11	LL1—LL3, LL9—LL11	LL1—LL3, LL9—LL11
LL1—LL3, LL7—LL11	LL1—LL3, LL7—LL11	LL1,LL3, LL8—LL11	LL1—LL5, LL7—LL11	LL1—LL5, LL7—LL11	LL1—LL3, LL7—LL11	LL1—LL3, LL7—LL11
LL1—LL4, LL6,LL7, LL9—LL11	LL1—LL3, LL7—LL11	LL1,LL2, LL4,LL9, LL10	LL1,LL5, LL9,LL10	LL1,LL2, LL4,LL5, LL9,LL10	LL1,LL2, LL11	LL1, LL9,LL10
LL1—LL5, LL9—LL11	LL1—LL11	LL1,LL5, LL9,LL10	LL1,LL5, LL9,LL10	LL1,LL5, LL9,LL10	LL1,LL2, LL11	LL1,LL5, LL9,LL10
LL1,LL2, LL9,LL11	LL1,LL9, LL11	LL1,LL2, LL6,LL9	LL1,LL5, LL9,LL10	LL1,LL2, LL4,LL5, LL9,LL10	LL1,LL11	LL1,LL9
LL1,LL8, LL9,LL11	LL1,LL8, LL9,LL11	LL1,LL8, LL9,LL11	LL1—LL11	LL1—LL5, LL7—LL11	LL1—LL3, LL7—LL11	LL1—LL3, LL7—LL11
LL1,LL2, LL7—LL9, LL11	LL1,LL8, LL9,LL11	LL1,LL2, LL7—LL9, LL11	LL1—LL5, LL7—LL11	LL1—LL11	LL1—LL3, LL7—LL11	LL1—LL3, LL7—LL11
LL1,LL2, LL10	LL1,LL2, LL10	LL1,LL10	LL1—LL5, LL9—LL11	LL1—LL5, LL9—LL11	LL1—LL4, LL6,LL7, LL9—LL11	LL1—LL5, LL9—LL11
LL1,LL9, LL11	LL1,LL8, LL9,LL11	LL1,LL9	LL1—LL5, LL9—LL11	LL1—LL5, LL9—LL11	LL1—LL3, LL7—LL11	LL1—LL11

表 2.19　$top(线,体) \wedge top(体,线)$

$\wedge$	$\begin{bmatrix}0&0&1\\0&0&1\\1&1&1\end{bmatrix}$ TL1	$\begin{bmatrix}0&0&1\\0&1&1\\1&1&1\end{bmatrix}$ TL2	$\begin{bmatrix}0&0&1\\1&0&1\\1&1&1\end{bmatrix}$ TL3	$\begin{bmatrix}0&0&1\\1&1&1\\1&1&1\end{bmatrix}$ TL4	$\begin{bmatrix}0&0&1\\1&1&1\\0&0&1\end{bmatrix}$ TL5
$\begin{bmatrix}0&0&1\\0&0&1\\1&1&1\end{bmatrix}$ LT1	LL1—LL11	LL1—LL5, LL9—LL11	LL1—LL5, LL9—LL11	LL1—LL5, LL9—LL11	LL1
$\begin{bmatrix}0&0&1\\0&1&1\\1&1&1\end{bmatrix}$ LT2	LL1—LL3, LL7—LL11	LL1—LL4, LL6, LL7, LL9—LL11	LL1—LL5, LL9—LL11	LL1—LL5, LL9—LL11	LL1, LL2, LL10
$\begin{bmatrix}0&0&1\\0&0&1\\1&1&1\end{bmatrix}$ LT3	LL1—LL3, LL7—LL11	LL1—LL3, LL7—LL11	LL1—LL11	LL1—LL5, LL7—LL11	LL1, LL8, LL9, LL11
$\begin{bmatrix}0&0&1\\0&1&1\\1&1&1\end{bmatrix}$ LT4	LL1—LL3, LL7—LL11	LL1—LL3, LL7—LL11	LL1—LL5, LL9—LL11	LL1—LL11	LL1—LL3, LL7—LL11
$\begin{bmatrix}0&1&0\\0&1&0\\1&1&1\end{bmatrix}$ LT5	LL1	LL1, LL2, LL11	LL1, LL5, LL9, LL10	LL1—LL5, LL9—LL11	LL1—LL11
$\begin{bmatrix}1&1&1\\1&0&1\\1&1&1\end{bmatrix}$ LT6	LL1—LL3, LL7—LL11	LL1—LL3, LL7—LL11	LL1—LL3, LL7—LL11	LL1—LL3, LL7—LL11	LL1, LL8, LL9, LL11
$\begin{bmatrix}1&1&1\\0&1&1\\1&1&1\end{bmatrix}$ LT7	LL1—LL3, LL7—LL11	LL1—LL3, LL7—LL11	LL1—LL3, LL7—LL11	LL1—LL3, LL7—LL11	LL1—LL3, LL7—LL11
$\begin{bmatrix}1&1&1\\0&0&1\\1&1&1\end{bmatrix}$ LT8	LL1—LL3, LL7—LL11	LL1—LL3, LL7—LL11	LL1—LL3, LL7—LL11	LL1—LL3, LL7—LL11	LL1, LL8, LL9, LL11
$\begin{bmatrix}1&0&0\\0&1&0\\1&1&1\end{bmatrix}$ LT9	LL1	LL1, LL2	LL1, LL10	LL1, LL2, LL10	LL1, LL2, LL10
$\begin{bmatrix}1&0&0\\1&1&0\\1&1&1\end{bmatrix}$ LT10	LL1	LL1, LL2	LL1, LL10	LL1, LL2, LL10	LL1, LL2, LL10
$\begin{bmatrix}1&0&0\\1&0&0\\1&1&1\end{bmatrix}$ LT11	LL1	LL1	LL1	LL1	LL1

第 2 章　拓扑关系描述及其定性推理　　55

→ *top*(线,线)

$\begin{bmatrix}1&1&1\\1&0&1\\1&1&1\end{bmatrix}$	$\begin{bmatrix}1&0&1\\1&1&1\\1&1&1\end{bmatrix}$	$\begin{bmatrix}1&0&1\\1&0&1\\1&1&1\end{bmatrix}$	$\begin{bmatrix}1&0&1\\0&1&1\\0&0&1\end{bmatrix}$	$\begin{bmatrix}1&1&1\\0&1&1\\0&0&1\end{bmatrix}$	$\begin{bmatrix}1&1&1\\0&0&1\\0&0&1\end{bmatrix}$
TL6	TL7	TL8	TL9	TL10	TL11
LL1—LL5, LL9—LL11	LL1—LL5, LL9—LL11	LL1—LL5, LL9—LL11	LL1	LL1	LL1
LL1—LL5, LL9—LL11	LL1—LL5, LL9—LL11	LL1—LL5, LL9—LL11	LL1,LL2	LL1,LL2	LL1
LL1—LL5, LL9—LL11	LL1—LL5, LL9—LL11	LL1—LL5, LL9—LL11	LL1,LL11	LL1,LL11	LL1
LL1—LL5, LL9—LL11	LL1—LL5, LL9—LL11	LL1—LL5, LL9—LL11	LL1,LL2, LL11	LL1,LL2, LL11	LL1
LL1,LL5, LL9,LL10	LL1—LL5, LL9—LL11	LL1,LL5, LL9,LL10	LL1,LL2, LL11	LL1,LL2, LL11	LL1
LL1—LL11	LL1—LL5, LL7—LL11	LL1—LL5, LL7—LL11	LL1,LL3, LL8—LL11	LL1—LL3, LL7—LL11	LL1—LL3, LL7—LL11
LL1—LL5, LL7—LL11	LL1—LL11	LL1—LL5, LL7—LL11	LL1,LL2, LL7—LL9, LL11	LL1,LL2, LL7—LL9, LL11	LL1,LL8, LL9,LL11
LL1—LL5, LL7—LL11	LL1—LL5, LL7—LL11	LL1—LL11	LL1,LL8, LL9,LL11	LL1,LL8, LL9,LL11	LL1,LL8, LL9,LL11
LL1,LL3, LL5, LL9—LL11	LL1,LL2, LL4,LL5, LL9,LL10	LL1,LL9, LL10	LL1,LL2, LL6,LL9	LL1,LL2, LL9,LL11	LL1,LL9, LL11
LL1—LL5, LL9—LL11	LL1,LL2, LL4,LL5, LL9,LL10	LL1,LL5, LL9,LL10	LL1,LL2, LL10	LL1—LL4, LL6,LL7, LL9—LL11	LL1—LL3, LL7—LL11
LL1—LL5, LL9—LL11	LL1,LL5, LL9,LL10	LL1,LL5, LL9,LL10	LL1,LL5, LL9,LL10	LL1—LL5, LL9—LL11	LL1—LL11

表 2.20　$top(点,B) \wedge top(B,点) \rightarrow top(点,点)$

$\wedge$	$C \in B°$	$C \in \partial B$	$C \in B^-$
$A \in B°$	d,e	d	d
$A \in \partial B$	d	d,e	d
$A \in B^-$	d	d	d,e

第3章 方向关系描述及其定性推理

在日常生活中,人们常说"在……北面,在……东面,……",人们在不经意间使用了方向关系:北、东等。在 GIS 中,方向关系描述了一个物体与另一个物体之间的相对方位关系,表示两个空间物体间的一种空间顺序关系。方向关系经常作为空间查询的选取条件和影像相似性评估的标准,是 GIS 分析中的重要研究内容。按照研究内容,GIS 中方向关系研究可分为方向关系的描述研究和方向关系的定性推理研究;按照研究范围,方向关系可分为 2D GIS 中的方向关系和 3D GIS 中的方向关系。目前,人们对方向关系的研究集中于二维空间方向关系的描述和推理。其中,方向关系的描述方法主要有两种:基于锥形的方向关系描述模型(Peuquet et al,1987)和基于投影的方向关系描述模型(Frank,1992,1996)。方向关系定性推理常用方法有:代数法(Frank,1991,1996),和根据 Allen 的区间关系推导(Freksa et al,1993)。本章的主要研究内容是采用投影法建立三维空间方向关系的描述模型,用代数法研究方向关系定性推理。

§3.1 二维方向关系描述及其定性推理

3.1.1 二维方向关系描述

方向关系描述的一个重要内容就是方向区域的划分。2D GIS 中方向区域的划分方法主要有两种:锥形方法(Peuquet et al,1987)(图 3.1)和投影方法(Frank,1992,1996)(图 3.2)。由于锥形方法没有考虑参照物的大小和形状对方向划分的影响,因此,当参照物为非点状物体时,采用锥形方法具有一定的局限性,故通常采用投影方法划分方向区域。方向区域的语意分辨率决定了划分方向区域的个数。经常使用的方向区域有四方向区域(图 3.2(a)),即东(O_E)、南(O_S)、西(O_W)、北(O_N) 4 个方向区域和八方向区域(图 3.2(b)),即东(O_E)、南(O_S)、西(O_W)、北(O_N)、东南(O_{SE})、西南(O_{SW})、东北(O_{NE})和西北(O_{NW})。对于非点状参照物,采用投影方法划分参照物所在的空间,将其划分为 8 个方向区域和一个 same 区域(O_{same}),如图 3.3 所示。

空间中任一物体与参照物的方向关系常用九交矩阵描述(Goyal,2001),即

$$\begin{bmatrix} O_{NW} \cap B & O_N \cap B & O_{NE} \cap B \\ O_W \cap B & O_{same} \cap B & O_E \cap B \\ O_{SW} \cap B & O_S \cap B & O_{SE} \cap B \end{bmatrix} \quad (3.1)$$

式中，O_i，$i \in \{N, NE, E, SE, S, SW, W, NW, same\}$是原子方向。

北
西
东
南
北
西北
东北
西
东
西南
东南
南

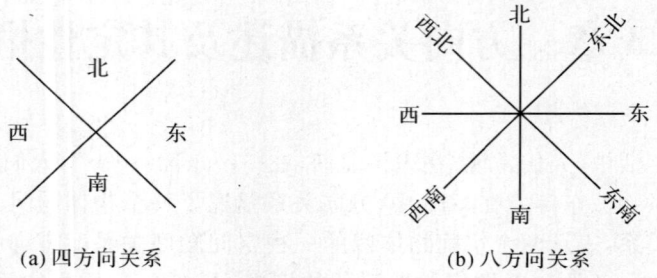

(a) 四方向关系　　(b) 八方向关系

图 3.1　基于锥形方法的方向关系

西
东
北
南
北
西北
东北
西
东
西南
东南
南

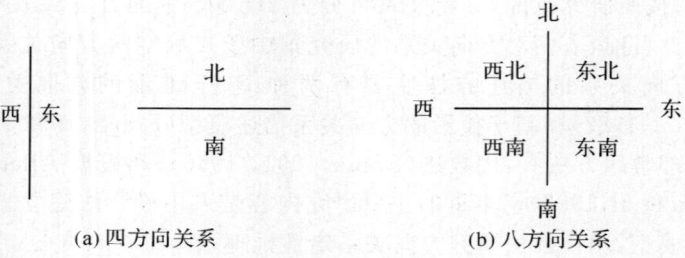

(a) 四方向关系　　(b) 八方向关系

图 3.2　基于投影的方向关系

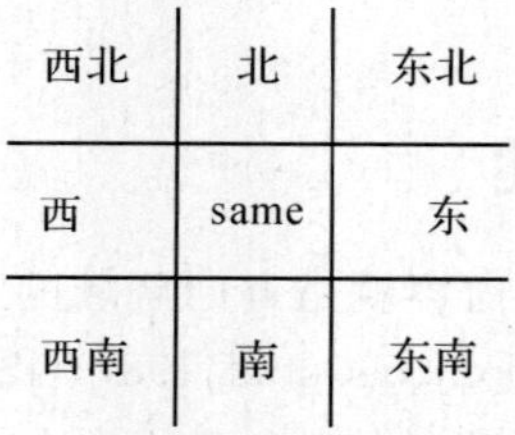

西北
北
东北
西
same
东
西南
南
东南

图 3.3　带中心区的八方向关系

当计算方向关系的隶属函数为下列布尔函数时，称九交矩阵式(3.1)为粗略方向关系矩阵，也称为 0-1 矩阵。

$$f(O_i, B)=\begin{cases}1 & O_i \cap B \neq \varnothing \\ 0 & O_i \cap B = \varnothing\end{cases} \tag{3.2}$$

式中，O_i 为原子方向。

当隶属函数为下列几何度量值的比值

$$f(O_i \cap B)=\frac{area(O_i \cap B)}{area(B)} \tag{3.3}$$

称下列九交矩阵表达式为详细方向关系矩阵

$$\begin{bmatrix} \dfrac{area(O_{\text{NW}} \bigcap B)}{area(B)} & \dfrac{area(O_{\text{N}} \bigcap B)}{area(B)} & \dfrac{area(O_{\text{NE}} \bigcap B)}{area(B)} \\ \dfrac{area(O_{\text{W}} \bigcap B)}{area(B)} & \dfrac{area(O_{\text{same}} \bigcap B)}{area(B)} & \dfrac{area(O_{\text{E}} \bigcap B)}{area(B)} \\ \dfrac{area(O_{\text{SW}} \bigcap B)}{area(B)} & \dfrac{area(O_{\text{S}} \bigcap B)}{area(B)} & \dfrac{area(O_{\text{SE}} \bigcap B)}{area(B)} \end{bmatrix} \quad (3.4)$$

式中，area 为几何度量算子。若目标对象为面状对象时，area 表示计算面积；如果目标对象为线状对象时，area 表示线的长度。

目标对象与参照对象之间的方向关系是由方向区域与目标对象相交的所有原子方向组成的集合，用 $dir(A,B) = \{O_i \mid f(O_i \bigcap B) = 1, i \in \{\text{N, NE, E, SE, S, SW, W, NW, same}\}\}$ 表示八方向关系系统中目标对象 B 与参照物 A 的方向关系。

3.1.2 二维方向关系推理

2D GIS 中方向关系推理有基于一个参照系的推理和基于两个参照系的推理(杜世宏，2004)，其中，基于一个参照系的推理是根据参照物 A 和 B 以及 A 和 C 间的方向关系，推导 B 和 C 间的方向关系；基于两个参照系的推理是根据参照物 A 与 B 的方向关系以及参照物 B 与 C 间的方向关系推导参照物 A 与 C 的方向关系。按照方向关系的多少，方向关系推理可分为：单方向关系与单方向关系推理(简称单方向关系推理)、单方向关系与多方向关系推理、多方向关系与单方向关系推理以及多方向关系与多方向关系推理。单方向关系是方向关系定性推理的基础，其他类型的方向关系推理可以根据单方向关系的组合运算求得。单方向关系推理方法主要有 3 种：①根据方向关系的定义和格网阵列的组合运算(曹菡 等，2001b)；②根据 Allen 的区间关系推导(Allen，1983；Freksa et al，1993)；③代数推理方法(Frank，1991，1996)。在表示单方向关系推理的结果时，方向关系矩阵通常用格网阵列表示(曹菡 等，2001b)，或用图标表示(Sharma，1996)。

§3.2 三维方向关系描述及其定性推理

3.2.1 三维方向关系描述

本书采用单纯形数据模型描述三维空间实体，通过研究单纯形间的方向关系研究空间实体间的方向关系。

设单纯形 A 为参照物，划分参照物 A 所在空间的函数为(刘新 等，2007b)：

$\pi_{X_{\max}}: x_{\max} = \max\{x_A \mid (x_A, y_A, z_A) \in A\}, \pi_{X_{\min}}: x_{\min} = \min\{x_A \mid (x_A, y_A, z_A) \in A\}$

$\pi_{Y_{\max}}: y_{\max} = \max\{y_A \mid (x_A, y_A, z_A) \in A\}, \pi_{Y_{\min}}: y_{\min} = \min\{y_A \mid (x_A, y_A, z_A) \in A\}$

$\pi_{Z_{\max}}:z_{\max}=\max\{z_A\mid(x_A,y_A,z_A)\in A\}, \pi_{Z_{\min}}:z_{\min}=\min\{z_A\mid(x_A,y_A,z_A)\in A\}$

其中,平面 $\pi_{Z_{\max}}$ 和 $\pi_{Z_{\min}}$ 将参照物所在的空间划分为 3 个部分(图 3.4),即

$$\text{up}=\{(x,y,z)\mid z>z_{\max}\} \tag{3.5}$$

$$\text{between}=\{(x,y,z)\mid z_{\min}\leqslant z\leqslant z_{\max}\} \tag{3.6}$$

$$\text{down}=\{(x,y,z)\mid z<z_{\min}\} \tag{3.7}$$

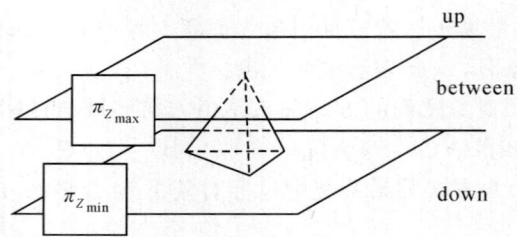

图 3.4 平面 $\pi_{Z_{\max}}$ 和 $\pi_{Z_{\min}}$ 划分的空间

平面 $\pi_{X_{\max}}$, $\pi_{X_{\min}}$, $\pi_{Y_{\max}}$ 和 $\pi_{Y_{\min}}$ 将参照物所在的空间划分为 9 个部分(图 3.5),即

$$\text{N}=\{(x,y,z)\mid x_{\min}\leqslant x\leqslant x_{\max},y>y_{\max}\} \tag{3.8}$$

$$\text{NE}=\{(x,y,z)\mid x>x_{\max},y>y_{\max}\} \tag{3.9}$$

$$\text{E}=\{(x,y,z)\mid x>x_{\max},y_{\min}\leqslant y\leqslant y_{\max}\} \tag{3.10}$$

$$\text{SE}=\{(x,y,z)\mid x>x_{\max},y<y_{\min}\} \tag{3.11}$$

$$\text{S}=\{(x,y,z)\mid x_{\min}\leqslant x\leqslant x_{\max},y<y_{\min}\} \tag{3.12}$$

$$\text{SW}=\{(x,y,z)\mid x<x_{\min},y<y_{\min}\} \tag{3.13}$$

$$\text{W}=\{(x,y,z)\mid x<x_{\min},y_{\min}\leqslant y\leqslant y_{\max}\} \tag{3.14}$$

$$\text{NW}=\{(x,y,z)\mid x<x_{\min},y>y_{\max}\} \tag{3.15}$$

$$\text{same}=\{(x,y,z)\mid x_{\min}\leqslant x\leqslant x_{\max},y_{\min}\leqslant y\leqslant y_{\max}\} \tag{3.16}$$

NW	N	NE
W	same	E
SW	S	SE

图 3.5 平面 $\pi_{X_{\max}}$, $\pi_{X_{\min}}$, $\pi_{Y_{\max}}$ 和 $\pi_{Y_{\min}}$ 和平面 XOY 的交所形成的图形

因此,六个划分平面($\pi_{X_{\max}}$, $\pi_{X_{\min}}$, $\pi_{Y_{\max}}$, $\pi_{Y_{\min}}$, $\pi_{Z_{\max}}$ 和 $\pi_{Z_{\min}}$)将参照物所在的空间划分为 $3\times9=27$ 个部分 $O_{i,j}$, $i\in\{\text{N,NE,E,SE,S,SW,W,NW,same}\}$, $j\in\{\text{up, between, down}\}$,如图 3.6 所示。

第 3 章 方向关系描述及其定性推理

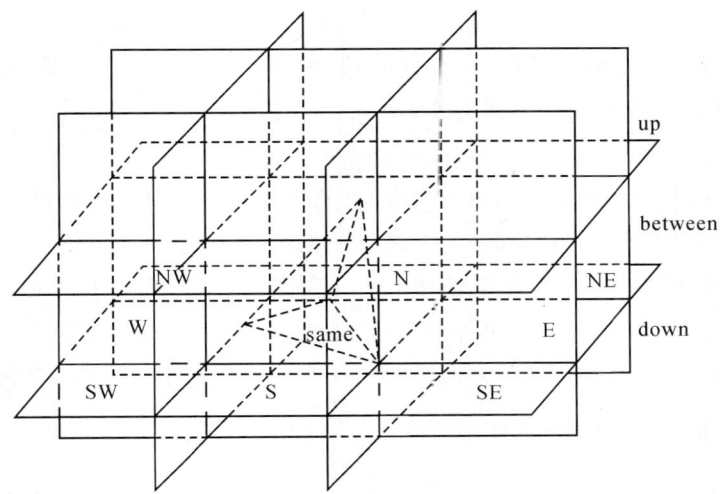

图 3.6 平面 $\pi_{X_{\max}}, \pi_{X_{\min}}, \pi_{Y_{\max}}, \pi_{Y_{\min}}, \pi_{Z_{\max}}$ 和 $\pi_{Z_{\min}}$ 划分后的空间

由于 1-单纯形、2-单纯形和 3-单纯形是凸的,所以过参照物在坐标轴上投影的最大(小)点的划分平面,实际上是过描述单纯形的节点在坐标轴上投影的最大(小)值的平面。例如,$x_{\max}=\max\{x_A\mid(x_A,y_A,z_A)\in A\}=\max\{x_i\mid(x_i,y_i,z_i),i=1,\cdots,n\}$,当 A 为 0-单纯形时,$n=1$;当 A 为 1-单纯形时,$n=2$;A 为 2-单纯形时,$n=3$;当 A 为 3-单纯形时,$n=4$。其他划分函数类似。

由上述方向区域的定义可知,当参照物为点时

$$\max\{x_A\mid(x_A,y_A,z_A)\in A\}=\min\{x_A\mid(x_A,y_A,z_A)\in A\} \tag{3.17}$$

$$\max\{y_A\mid(x_A,y_A,z_A)\in A\}=\min\{y_A\mid(x_A,y_A,z_A)\in A\} \tag{3.18}$$

$$\max\{z_A\mid(x_A,y_A,z_A)\in A\}=\min\{z_A\mid(x_A,y_A,z_A)\in A\} \tag{3.19}$$

所以,参照物为点时的方向区域划分函数是参照物为体时划分函数的特例。

由于隶属函数

$$f(O_{i,j},B)=\begin{cases}1 & B\cap O_{i,j}\neq\varnothing\\ 0 & B\cap O_{i,j}=\varnothing\end{cases} \tag{3.20}$$

式中,$i\in DIRXY=\{\text{N, NE, E, SE, S, SW, W, NW, same}\}$,$j\in DIRZ=\{\text{up, between, down}\}$,只刻画了目标对象与方向区域是否相交,而无法描述交在量上的变化。所以,为提高隶属函数的描述能力,采用隶属函数

$$f(O_{i,j}\cap B)=\frac{area(O_{i,j}\cap B)}{area(B)} \tag{3.21}$$

式中,$i\in\{\text{N, NE, E, SE, S, SW, W, NW, same}\}$,$j\in\{\text{up, between, down}\}$。当目标

对象为线段时,算子 $area$ 计算线段的长度;当目标对象为三角形时,算子 $area$ 计算三角形的面积;当目标对象为四面体时,算子 $area$ 计算四面体的体积。因此,目标对象 B 与参照物 A 间的方向关系为

$$dir(A,B) = \{f(O_{i,j} \cap B)/O_{i,j} \mid f(O_{i,j} \cap B) > 0\} \quad (3.22)$$

在基于投影的方向关系描述模型中,参照物确定了方向区域的划分,同一个目标对象相对于不同的参照物其方向关系是不同的。为此,在同一个研究空间中存在多个参照物的情况下,用方向区域 $O_{i,j}$,$i \in \{N_A, NE_A, E_A, SE_A, S_A, SW_A, W_A, NW_A, same_A\}$,$j \in \{up_A, between_A, down_A\}$ 表示由参照物 A 确定的方向区域。方向区域 $O_{i,j}$ 是由 X-Y 轴确定的方向区域 i 和 Z 轴确定的方向区域 j 共同确定的。

3.2.2 三维方向关系推理

在 3D GIS 中,按照目标对象与方向区域相交个数的多少,可以将方向关系分为单方向关系和多方向关系。单方向关系是目标对象只与一个方向区域相交时的方向关系,多方向关系是目标对象与多个方向区域相交时的方向关系。本书运用定义法和代数法研究方向关系推理。对于单方向关系推理,给出单方向关系的组合推理表。为形象直观地描述空间物体间的方向关系,用图标描述物体间的方向关系,如图 3.7 所示。用图 3.8(a) 描述图 3.3 所示的方向关系,用图 3.8(b) 描述三维空间中任意两个物体间的方向关系。

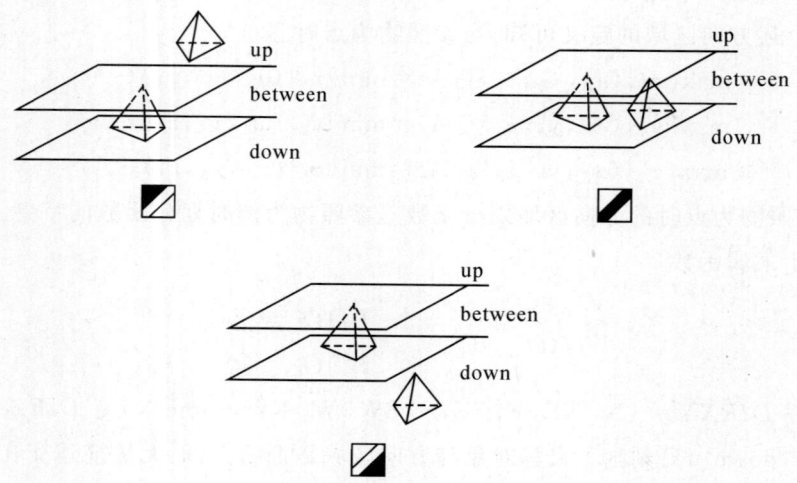

图 3.7 目标对象在参照物的上、中、下三个位置所对应的图

当目标对象为点时,方向关系为单方向关系。另外,当目标对象为 1-单纯形、2-单纯形或 3-单纯形时,可能为单方向关系,也可能为多方向关系。

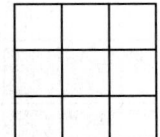 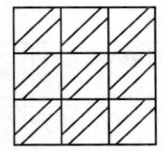

(a) 二维空间方向关系的描述图标　　(b) 三维空间方向关系的描述图标

图 3.8　描述空间方向关系的图标

设 A、B、C 是三维空间中的三个物体，方向关系 $dir(A,B)$，$dir(B,C)$ 分别为 A、B 间的方向关系和 B、C 间的方向关系，可以根据 $dir(A,B)$ 和 $dir(B,C)$，推断 $dir(A,C)$，表示为

$$dir(A,B) \wedge dir(B,C) \rightarrow dir(A,C) \tag{3.23}$$

根据三维空间中 27 个方向区域的定义和代数法进行方向关系推理，如例 3.1 所示。

例 3.1

$$O_{N,up} \wedge O_{NE,between} = \{(x_B, y_B, z_B) \mid \min_{(x_A, y_A, z_A) \in A} \{x_A\} \leqslant x_B \leqslant \max_{(x_A, y_A, z_A) \in A} \{x_A\}, y_B >$$
$$\max_{(x_A, y_A, z_A) \in A} \{y_A\}, z_B > \max_{(x_A, y_A, z_A) \in A} \{z_A\}\} \bigcap \{(x_C, y_C, z_C) \mid x_C >$$
$$\max_{(x_B, y_B, z_B) \in B} \{x_B\}, y_C > \max_{(x_B, y_B, z_B) \in B} \{y_B\}, \min_{(x_B, y_B, z_B) \in B} \{z_B\} \leqslant z_C \leqslant$$
$$\max_{(x_B, y_B, z_B) \in B} \{z_B\}\}$$
$$\rightarrow \{(x_C, y_C, z_C) \mid \min_{(x_A, y_A, z_A) \in A} \{x_A\} \leqslant x_C \leqslant \max_{(x_A, y_A, z_A) \in A} \{x_A\} \text{ or } x_C >$$
$$\max_{(x_A, y_A, z_A) \in A} \{x_A\}, y_C > \max_{(x_A, y_A, z_A) \in A} \{y_A\}, z_C > \max_{(x_A, y_A, z_A) \in A} \{z_A\}\}$$
$$\rightarrow O_{N,up} \vee O_{NE,up}$$

例 3.1 的推理结果表明，当 $dir(A,B) = O_{N,up}$，$dir(B,C) = O_{NE,between}$ 时，$dir(A,C)$ 有三种可能：①$dir(A,C) = O_{NE,up}$，②$dir(A,C) = O_{N,up}$，③$dir(A,C) = O_{N,up}$ and $O_{NE,up}$，用表 3.1 表示推理 $O_{N,up} \wedge O_{NE,between} \rightarrow O_{N,up} \vee O_{NE,up}$。

表 3.1　$O_{N,up} \wedge O_{NE,between}$ 推理组合

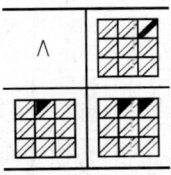

在三维空间中，单方向关系推理共有 $27 \times 27 = 729$ 个组合推理，推理结果见表 3.2。

64 三维空间关系的描述及其定性推理

表 3.2 三维空间

第 3 章 方向关系描述及其定性推理

单方向推理组合

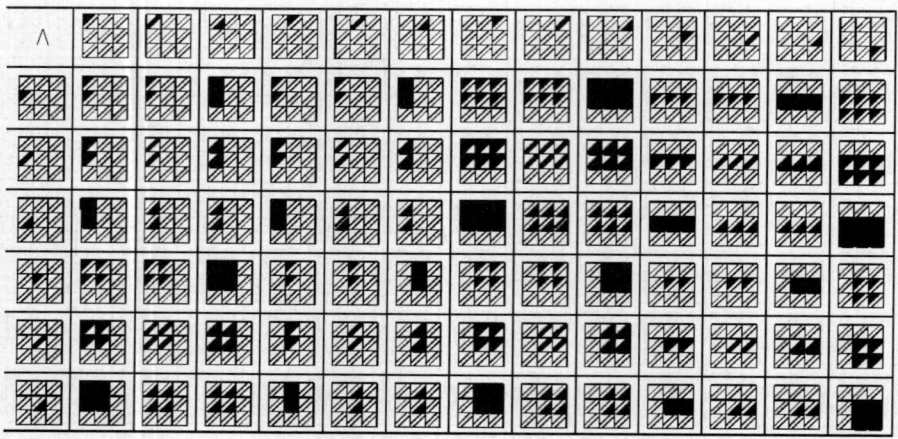

当 A、B 为单方向关系, B、C 为多方向关系时, 可以根据单方向关系的推理结果, 进行方向关系推理。设 $dir(A,B)=\{dir_{AB}\}$, $dir(B,C)=\{dir_{1BC},dir_{2BC},\cdots,dir_{nBC}\}$, 方向关系推理的方法为

$$dir(A,B) \wedge dir(B,C) = \{dir_{AB}\} \wedge \{dir_{1BC}, dir_{2BC}, \cdots, dir_{nBC}\} = (dir_{AB} \wedge dir_{1BC}) \vee \\ (dir_{AB} \wedge dir_{2BC}) \vee \cdots \vee (dir_{AB} \wedge dir_{nBC}) \quad (3.24)$$

例 3.2 设 $dir(A,B)=O_{\text{N,up}}$, $dir(B,C)=\{O_{\text{NE,between}}, O_{\text{NE,down}}\}$, 求 $dir(A,B) \wedge dir(B,C)$。

解: 由式(3.24)可知

$$dir(A,B) \wedge dir(B,C) = O_{\text{N,up}} \wedge \{O_{\text{NE,between}}, O_{\text{NE,down}}\} \\ = (O_{\text{N,up}} \wedge O_{\text{NE,between}}) \vee (O_{\text{N,up}} \wedge O_{\text{NE,down}})$$

根据方向区域的定义可知

$$O_{\text{N,up}} \wedge O_{\text{NE,down}} = \{(x_B, y_B, z_B) \mid \min_{(x_A,y_A,z_A) \in A}\{x_A\} \leqslant x_B \leqslant \max_{(x_A,y_A,z_A) \in A}\{x_A\}, \ y_B > \\ \max_{(x_A,y_A,z_A) \in A}\{y_A\}, z_B > \max_{(x_A,y_A,z_A) \in A}\{z_A\}\} \cap \{(x_C, y_C, z_C) \mid x_C > \\ \max_{(x_B,y_B,z_B) \in B}\{x_B\}, y_C > \max_{(x_B,y_B,z_B) \in B}\{y_B\}, z_C < \min_{(x_B,y_B,z_B) \in B}\{z_B\}\} \\ \rightarrow \{(x_C,y_C,z_C) \mid \min_{(x_A,y_A,z_A) \in A}\{x_A\} \leqslant x_C \leqslant \max_{(x_A,y_A,z_A) \in A}\{x_A\} \text{ or } x_C > \\ \max_{(x_A,y_A,z_A) \in A}\{x_A\}, y_C > \max_{(x_A,y_A,z_A) \in A}\{y_A\}, z_C > \max_{(x_A,y_A,z_A) \in A}\{z_A\} \text{ or } \\ \min_{(x_A,y_A,z_A) \in A}\{z_A\} \leqslant z_C \leqslant \max_{(x_A,y_A,z_A) \in A}\{z_A\} \text{ or } z_C < \min_{(x_A,y_A,z_A) \in A}\{z_A\}\} \\ \rightarrow O_{\text{N,up}} \vee O_{\text{NE,up}} \vee O_{\text{N,between}} \vee O_{\text{NE,between}} \vee O_{\text{N,down}} \vee O_{\text{NE,down}}$$

所以, $O_{\text{N,up}} \wedge O_{\text{NE,down}} \rightarrow O_{\text{N,up}} \vee O_{\text{NE,up}} \vee O_{\text{N,between}} \vee O_{\text{NE,between}} \vee O_{\text{N,down}} \vee O_{\text{NE,down}}$。

续表

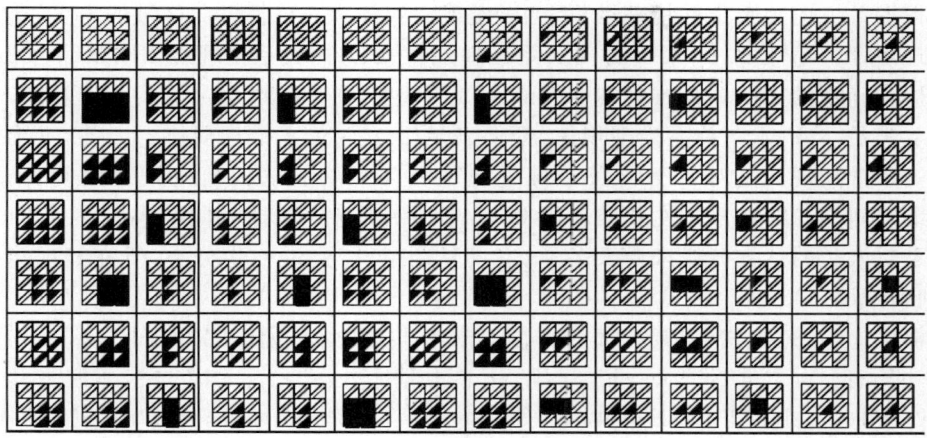

由例 3.1 可知,$O_{N,up} \wedge O_{NE,between} \to O_{N,up} \vee O_{NE,up}$。

$$O_{N,up} \wedge \{O_{NE,between}, O_{NE,down}\} = (O_{N,up} \wedge O_{NE,between}) \vee (O_{N,up} \wedge O_{NE,down})$$
$$\to O_{N,up} \vee O_{NE,up}) \vee (O_{N,up} \vee O_{NE,up} \vee O_{N,between} \vee O_{NE,between} \vee O_{N,down} \vee O_{NE,down})$$
$$\to O_{N,up} \vee O_{NE,up} \vee O_{N,between} \vee O_{NE,between} \vee O_{N,down} \vee O_{NE,down}$$

即 $O_{N,up} \wedge \{O_{NE,between}, O_{NE,down}\} \to O_{N,up} \vee O_{NE,up} \vee O_{N,between} \vee O_{NE,between} \vee O_{N,down} \vee O_{NE,down}$。

当 A、B 为多方向关系,B、C 为单方向关系时,可以根据单方向关系的推理结果,进行方向关系推理。设 $dir(A,B) = \{dir_{1AB}, dir_{2AB}, \cdots, dir_{nAB}\}$,$dir(B,C) = \{dir_{BC}\}$,方向关系推理的方法为

$$dir(A,B) \wedge dir(B,C) = \{dir_{1AB}, dir_{2AB}, \cdots, dir_{nAB}\} \wedge \{dir_{BC}\} = (dir_{1AB} \wedge dir_{BC}) \vee (dir_{2AB} \wedge dir_{BC}) \vee \cdots \vee (dir_{nAB} \wedge dir_{BC}) \quad (3.25)$$

当 A、B 间的方向关系和 B、C 间的方向关系均为多方向关系时,可以根据单方向和多方向关系进行推理。设 $dir(A,B) = \{dir_{1AB}, dir_{2AB}, \cdots, dir_{mAB}\}$,$dir(B,C) = \{dir_{1BC}, dir_{2BC}, \cdots, dir_{nBC}\}$,方向关系推理的方法为

$$\begin{aligned}dir(A,B) \wedge dir(B,C) &= \{dir_{1AB}, dir_{2AB}, \cdots, dir_{mAB}\} \wedge \{dir_{1BC}, dir_{2BC}, \cdots, dir_{nBC}\} \\ &= (dir_{1AB} \wedge \{dir_{1BC}, dir_{2BC}, \cdots, dir_{nBC}\}) \vee \cdots \vee (dir_{mAB} \wedge \{dir_{1BC}, dir_{2BC}, \cdots, dir_{nBC}\}) \\ &= ((dir_{1AB} \wedge dir_{1BC}) \vee (dir_{1AB} \wedge dir_{2BC}) \vee \cdots \vee (dir_{1AB} \wedge dir_{nBC})) \vee \cdots \vee ((dir_{mAB} \wedge dir_{1BC}) \vee (dir_{mAB} \wedge dir_{2BC}) \vee \cdots \vee (dir_{mAB} \wedge dir_{nBC}))\end{aligned} \quad (3.26)$$

3.2.3 单方向关系定性推理规律

定义 3.1 在三维空间中,称由 $X\text{-}Y$ 方向区域确定的关系为 $X\text{-}Y$ 方向关系。

定义 3.2 在三维空间中,称由 Z 方向区域确定的关系为 Z 方向关系。

根据单方向关系定性推理结果,可以发现一些规律。根据方向区域的定义,定性推理规律分为三类,即 X-Y 方向关系推理规律、Z 方向关系推理规律以及综合 3 个坐标轴确定的方向关系推理规律。综合 3 个坐标轴确定的方向关系推理规律,不仅要满足 X-Y 方向关系推理规律,还要满足 Z 方向关系推理规律。因此,三维方向关系推理规律是对 Z 方向关系推理规律和 X-Y 方向关系推理规律的复合。在三维空间中,X-Y 方向关系推理规律有 $9\times9=81$ 种(包括相同结果),然而 81 种关系对人类的认知分辨来讲太多了,也太复杂了,需要进一步分类,研究其规律性(刘新 等,2006)。

根据式(3.8)至式(3.15),部分方向区域有共同之处,如,X-Y 方向关系 N 与 NW、NE 有共同之处:$y>y_{\max}$;E 与 NE、SE 有共同之处:$x>x_{\max}$;NE 与 N、NW 有共同之处:$y>y_{\max}$;NE 与 E、SE 有共同之处:$x>x_{\max}$。在研究定性推理规律时,利用 X-Y 方向区域部分区域有共同之处的特点,将发现 X-Y 方向关系定性推理的规律。

在三维空间中,令 $\bigcup\limits_{k\in DIRZ} O_{i,k} = O_{i,\mathrm{up}} \vee O_{i,\mathrm{between}} \vee O_{i,\mathrm{down}}$, $j_1, j_2 \in \{\mathrm{up}, \mathrm{between}, \mathrm{down}\}$,$X$-$Y$ 方向关系定性推理规律为

$$O_{i,j_1} \wedge O_{i,j_2} \to \bigcup_{k\in DIRZ} O_{i,k} \tag{3.27}$$

式中,$i \in \{\mathrm{N, NE, E, SE, S, SW, W, NW, same}\}$。

$$O_{i,j_1} \wedge O_{\mathrm{same},j_2} \to \bigcup_{k\in DIRZ} O_{i,k} \tag{3.28}$$

式中,$i \in \{\mathrm{N, NE, E, SE, S, SW, W, NW, same}\}$。

$$O_{i,j_1} \wedge O_{\mathrm{N},j_2} \to \bigcup_{k\in DIRZ} O_{i,k} \tag{3.29}$$

式中,$i \in \{\mathrm{NW, N, NE}\}$。

$$O_{\mathrm{W},j_1} \wedge O_{\mathrm{N},j_2} \to \bigcup_{k\in DIRZ} (O_{\mathrm{W},k} \vee O_{\mathrm{NW},k}) \tag{3.30}$$

$$O_{\mathrm{same},j_1} \wedge O_{\mathrm{N},j_2} \to \bigcup_{k\in DIRZ} (O_{\mathrm{same},k} \vee O_{\mathrm{N},k}) \tag{3.31}$$

$$O_{\mathrm{E},j_1} \wedge O_{\mathrm{N},j_2} \to \bigcup_{k\in DIRZ} (O_{\mathrm{E},k} \vee O_{\mathrm{NE},k}) \tag{3.32}$$

$$O_{\mathrm{SW},j_1} \wedge O_{\mathrm{N},j_2} \to \bigcup_{k\in DIRZ} (O_{\mathrm{SW},k} \vee O_{\mathrm{W},k} \vee O_{\mathrm{NW},k}) \tag{3.33}$$

$$O_{\mathrm{S},j_1} \wedge O_{\mathrm{N},j_2} \to \bigcup_{k\in DIRZ} (O_{\mathrm{S},k} \vee O_{\mathrm{same},k} \vee O_{\mathrm{N},k}) \tag{3.34}$$

$$O_{\mathrm{SE},j_1} \wedge O_{\mathrm{N},j_2} \to \bigcup_{k\in DIRZ} (O_{\mathrm{SE},k} \vee O_{\mathrm{E},k} \vee O_{\mathrm{NE},k}) \tag{3.35}$$

$$O_{\mathrm{N},j_1} \wedge O_{\mathrm{NE},j_2} \to \bigcup_{k\in DIRZ} (O_{\mathrm{N},k} \vee O_{\mathrm{NE},k}) \tag{3.36}$$

$$O_{\mathrm{E},j_1} \wedge O_{\mathrm{NE},j_2} \to \bigcup_{k\in DIRZ} (O_{\mathrm{E},k} \vee O_{\mathrm{NE},k}) \tag{3.37}$$

$$O_{\mathrm{NW},j_1} \wedge O_{\mathrm{NE},j_2} \to \bigcup_{k\in DIRZ} (O_{\mathrm{NW},k} \vee O_{\mathrm{N},k} \vee O_{\mathrm{NE},k}) \tag{3.38}$$

$$O_{\mathrm{SE},j_1} \wedge O_{\mathrm{NE},j_2} \to \bigcup_{k\in DIRZ} (O_{\mathrm{SE},k} \vee O_{\mathrm{E},k} \vee O_{\mathrm{NE},k}) \tag{3.39}$$

$$O_{\text{same},j_1} \wedge O_{\text{NE},j_2} \to \bigcup_{k\in DIRZ}(O_{\text{same},k} \vee O_{\text{NE},k} \vee O_{\text{N},k} \vee O_{\text{E},k}) \tag{3.40}$$

$$O_{\text{SW},j_1} \wedge O_{\text{NE},j_2} \to \bigcup_{k\in DIRZ}(O_{\text{NW},k} \vee O_{\text{N},k} \vee O_{\text{NE},k} \vee O_{\text{W},k} \vee O_{\text{same},k} \vee$$
$$O_{\text{E},k} \vee O_{\text{SW},k} \vee O_{\text{S},k} \vee O_{\text{SE},k}) \tag{3.41}$$

$$O_{\text{SW},j_1} \wedge O_{\text{NE},j_2} = O_{\text{NE},j_1} \wedge O_{\text{SW},j_2} = O_{\text{SE},j_1} \wedge O_{\text{NW},j_2} = O_{\text{NW},j_1} \wedge O_{\text{SE},j_2} \tag{3.42}$$

根据式(3.27)可知,X-Y方向关系具有传递性。式(3.28)说明,当$dir(B,C)$的X-Y方向关系是 same 时,推理结果等于$dir(A,B)$的X-Y方向关系。式(3.29)说明,当$dir(B,C)$的X-Y方向关系为 N、E、S、W 时,如果$dir(A,B)$的X-Y方向关系与$dir(B,C)$的X-Y方向关系有共同的方向,$dir(A,C)$的X-Y方向关系等于$dir(A,B)$的X-Y方向关系,如表 3.3 和表 3.4 所示。表 3.3 说明:当B在参照物A的北部,即$i\in\{\text{NW},\text{N},\text{NE}\}$,如果$dir(B,C)$的$X$-$Y$方向关系为北,即 N,则$dir(A,C)$的$X$-$Y$方向关系等于$dir(A,B)$的$X$-$Y$方向关系。表 3.4 说明:当$B$在参照物$A$的东部,即$i\in\{\text{NE},\text{E},\text{SE}\}$,如果$dir(B,C)$的$X$-$Y$方向关系为东,即 E,则$dir(A,C)$的$X$-$Y$方向关系等于$dir(A,B)$的$X$-$Y$方向关系。

表 3.3 $O_{i,j_1} \wedge O_{\text{N},j_2}$ 推理组合

表 3.4 $O_{i,j_1} \wedge O_{\text{E},j_2}$ 推理组合

当 $dir(B,C)$ 的 X-Y 方向关系为 N、E、S、W 时，定性推理的规律即为式(3.30)至式(3.35)。平面 $\pi_{X_{\min}}$ 和 $\pi_{X_{\max}}$ 划分参照物 A 所在的空间成 3 个部分：$\{(x,y,z)|x<\min\{x_A|(x_A,y_A,z_A)\in A\}\}$(西部)、$\{(x,y,z)|\min\{x_A|(x_A,y_A,z_A)\in A\}\leqslant x\leqslant\min\{x_A|(x_A,y_A,z_A)\in A\}\}$(中部)和$\{(x,y,z)|x>\max\{x_A|(x_A,y_A,z_A)\in A\}\}$(东部)。当 $dir(B,C)$ 的 X-Y 方向区域为 N，$dir(A,B)$ 的 X-Y 方向区域分别为西部、中部和东部时，$dir(A,C)$ 的 X-Y 方向区域相应为：西部、中部和东部。表 3.5 和表 3.6 表示：当 $dir(B,C)$ 的 X-Y 方向区域为 N，$dir(A,B)$ 的 X-Y 方向区域为东部时，$dir(A,C)$ 的 X-Y 方向区域为东部。将式(3.30)至式(3.35)中 $dir(B,C)$ 的 X-Y 方向关系 N 换为 S，结果类似。

表 3.5 $O_{E,j_1} \wedge O_{N,j_2}$ 推理组合

表 3.6 $O_{SE,j_1} \wedge O_{N,j_2}$ 推理组合

式(3.36)至式(3.37)说明：当 $dir(B,C)$ 的 X-Y 方向区域为 SW、NE、NW 或 SE 时，如果 $dir(A,B)$ 的 X-Y 方向区域与 $dir(B,C)$ 的 X-Y 方向区域相邻，如 S 与 SW 相邻，W 与 SW 相邻，则 $dir(A,C)$ 的 X-Y 方向区域为 $dir(A,B)$ 的 X-Y 方向区域或 $dir(B,C)$ 的 X-Y 方向区域或两者的和，如表 3.7 所示。

表 3.7 $O_{S,j_1} \wedge O_{SW,j_2}$ 推理组合

式(3.38)至式(3.39)说明：当 $dir(A,B)$ 和 $dir(B,C)$ 的 X-Y 方向区域为 SW、NE、NW 或 SE，且有共同方向时，$dir(A,C)$ 的 X-Y 方向区域规律：$dir(A,B)$ 的 X-Y 方向区域、$dir(B,C)$ 的 X-Y 方向区域以及共同方向三者的组合。根据式(3.38)和式(3.39)，可得 $O_{NE,j_1} \wedge O_{SE,j_2} \rightarrow \bigcup_{k \in DIRZ}(O_{NE,k} \vee O_{E,k} \vee O_{SE,k})$，表示当 $dir(A,C)$ 的 X-Y 方向区域是 NE，$dir(B,C)$ 的 X-Y 方向区域是 SE，NE 与 SE 有共同方向 E，$dir(A,C)$ 的 X-Y 方向区域是 NE∪E∪SE，具体推理结果见表 3.8。

第 3 章　方向关系描述及其定性推理　　　　　　　　　　　　　　　　　　71

表 3.8　$O_{NE,j_1} \wedge O_{SE,j_2}$ 推理组合

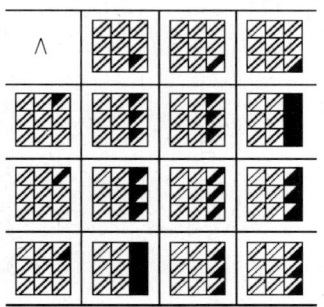

式(3.40)说明当 $dir(A,B)$ 的 X-Y 方向关系为 same，$dir(B,C)$ 的 X-Y 方向关系为 SW、NE、NW 或 SE 时，$dir(A,C)$ 的 X-Y 方向关系为 $dir(B,C)$ 的 X-Y 方向区域与其相邻区域的并。由式(3.40)可得

$$O_{same,j_1} \wedge O_{SW,j_2} \rightarrow \bigcup_{k \in DIRZ}(O_{same,k} \vee O_{SW,k} \vee O_{S,k} \vee O_{W,k}) \quad (3.43)$$

式(3.43)表示 $dir(A,B)$ 的 X-Y 方向区域为 same，$dir(B,C)$ 的 X-Y 方向区域为 SW，$dir(A,C)$ 的 X-Y 方向区域为 $dir(B,C)$ 的 X-Y 方向区域 SW 与其相邻区域(W，same，S)的并，具体推理结果如表 3.9 所示。

式(3.41)和式(3.42)说明：当 $dir(A,B)$ 和 $dir(B,C)$ 的 X-Y 方向关系是互为相反的关系 SW 和 NE，或 NW 和 SE 时，推理结果没有提供 X-Y 方向关系信息，推理结果如表 3.10 所示。

表 3.9　$O_{same,j_1} \wedge O_{SW,j_2}$ 推理组合　　　　表 3.10　$O_{SW,j_1} \wedge O_{NE,j_2}$ 推理组合

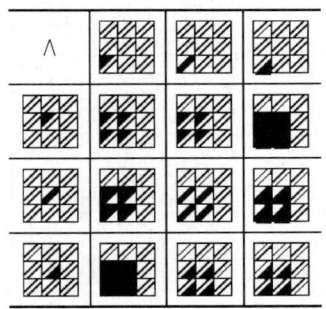 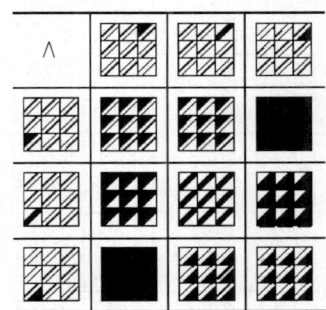

在三维空间中，Z 方向关系推理共有 3×3 个规律，将其归纳分类可得

$$O_{i_1,j} \wedge O_{i_2,j} \rightarrow O_{k_1,j} \vee O_{k_2,j} \vee \cdots \vee O_{k_l,j} \quad (3.44)$$

$$O_{i_1,j} \wedge O_{i_2,between} \rightarrow O_{k_1,j} \vee O_{k_2,j} \vee \cdots \vee O_{k_l,j} \quad (3.45)$$

$$O_{i_1,between} \wedge O_{i_2,j} \rightarrow O_{k_1,j} \vee O_{k_2,j} \vee \cdots \vee O_{k_l,j} \vee O_{k_1,between} \vee$$
$$O_{k_2,between} \vee \cdots \vee O_{k_l,between} \quad (3.46)$$

$$O_{i_1,up} \wedge O_{i_2,down} \rightarrow O_{k_1,up} \vee \cdots \vee O_{k_l,up} \vee O_{k_1,between} \vee \cdots \vee O_{k_l,between} \vee$$
$$O_{k_1,down} \vee \cdots \vee O_{k_l,down} \quad (3.47)$$

$$O_{i_1,up} \wedge O_{i_2,down} = O_{i_1,down} \wedge O_{i_2,up} \quad (3.48)$$

式(3.44)至式(3.48)中，$i_1,i_2,k_1,k_2 \cdots k_l \in \{N, NE, E, SE, S, SW, W, NW, same\}$，$j \in$

{up, between, down}。

证明:令 $DIRXY_A = \{N_A, NE_A, E_A, SE_A, S_A, SW_A, W_A, NW_A, same_A\}$, $DIRXY_B = \{N_B, NE_B, E_B, SE_B, S_B, SW_B, W_B, NW_B, same_B\}$。

——式(3.44)的证明如下。

式(3.44)规律共有以下 3 种情况

$$O_{i_1, up} \wedge O_{i_2, up} \rightarrow O_{k_1, up} \vee O_{k_2, up} \vee \cdots \vee O_{k_l, up} \tag{3.49}$$

$$O_{i_1, between} \wedge O_{i_2, between} \rightarrow O_{k_1, between} \vee O_{k_2, between} \vee \cdots \vee O_{k_l, between} \tag{3.50}$$

$$O_{i_1, down} \wedge O_{i_2, down} \rightarrow O_{k_1, down} \vee O_{k_2, down} \vee \cdots \vee O_{k_l, down} \tag{3.51}$$

其中,$k_1, k_2, \cdots, k_l \in DIRXY_A$ 由 $i_1 \in DIRXY_A$ 和 $i_2 \in DIRXY_B$ 确定。

$$O_{i_1, up} \wedge O_{i_2, up} = \{(x_B, y_B, z_B) | i_1 \in DIRXY_A, z_B > \max\{z_A | (x_A, y_A, z_A) \in A\}\} \cap$$
$$\{(x_C, y_C, z_C) | i_2 \in DIRXY_B, z_C > \max\{z_B | (x_B, y_B, z_B) \in B\}\}$$
$$\rightarrow \{(x_C, y_C, z_C) | k \in DIRXY_A, z_C > \max\{z_A | (x_A, y_A, z_A) \in A\}\}$$
$$\rightarrow O_{k_1, up} \vee O_{k_2, up} \vee \cdots \vee O_{k_l, up}$$

定性推理 $O_{i_1, up} \wedge O_{i_2, up}$ 的结果见表 3.11。

表 3.11 $O_{i_1, up} \wedge O_{i_2, up}$ 推理组合

$O_{i_1,\text{between}} \wedge O_{i_2,\text{between}} = \{(x_B, y_B, z_B) \mid i_1 \in DIRXY_A, \min\{z_A \mid (x_A, y_A, z_A) \in A\} \leqslant z_B \leqslant \max\{z_A \mid (x_A, y_A, z_A) \in A\}\} \cap \{(x_C, y_C, z_C) \mid i_2 \in DIRXY_B, \min\{z_B \mid (x_B, y_B, z_B) \in B\} \leqslant z_C \leqslant \max\{z_B \mid (x_B, y_B, z_B) \in B\}\}$

$\rightarrow \{(x_C, y_C, z_C) \mid k \in DIRXY_A, \min\{z_A \mid (x_A, y_A, z_A) \in A\} \leqslant z_C \leqslant \max\{z_A \mid (x_A, y_A, z_A) \in A\}\}$

$\rightarrow O_{k_1,\text{between}} \cup O_{k_2,\text{between}} \cup \cdots \cup O_{k_l,\text{between}}$

定性推理 $O_{i_1,\text{between}} \wedge O_{i_2,\text{between}}$ 的结果见表 3.12。

表 3.12 $O_{i_1,\text{between}} \wedge O_{i_2,\text{between}}$ 推理组合

$O_{i_1,\text{down}} \wedge O_{i_2,\text{down}} = \{(x_B, y_B, z_B) \mid i_1 \in DIRXY_A, z_B < \min\{z_A \mid (x_A, y_A, z_A) \in A\}\} \cap \{(x_C, y_C, z_C) \mid i_2 \in DIRXY_B, z_C < \min\{z_B \mid (x_B, y_B, z_B) \in B\}\}$

$\rightarrow \{(x_C, y_C, z_C) \mid k \in DIRXY_A, z_C < \min\{z_A \mid (x_A, y_A, z_A) \in A\}\}$

$\rightarrow O_{k_1,\text{down}} \cup O_{k_2,\text{down}} \cup \cdots \cup O_{k_l,\text{down}}$

定性推理 $O_{i,\text{down}} \wedge O_{j,\text{down}}$ 的结果见表 3.13。

表 3.13 $O_{i_1,\text{down}} \wedge O_{i_2,\text{down}}$ 推理组合

——式(3.45)的证明如下。

$O_{i_1,j} \wedge O_{i_2,\text{between}} = \{i_1 \in DIRXY_A, j \in DIRZ_A\} \cap \{i_2 \in DIRXY_B,$
$$\min_{(x_B,y_B,z_B) \in B} \{z_B\} \leqslant z_C \leqslant \max_{(x_B,y_B,z_B) \in B} \{z_B\}\} \tag{3.52}$$

(1) 如果 $j = \text{up}_A$,即 $z_B > \max\limits_{(x_A,y_A,z_A) \in A} \{z_A\}$,则 $\min\limits_{(x_B,y_B,z_B) \in B} \{z_B\} > \max\limits_{(x_A,y_A,z_A) \in A} \{z_A\}$,再由式(3.52)得 $z_C > \max\limits_{(x_A,y_A,z_A) \in A} \{z_A\}$。

(2) 如果 $j = \text{between}_A$,即 $\min\limits_{(x_A,y_A,z_A) \in A} \{z_A\} \leqslant z_B \leqslant \max\limits_{(x_A,y_A,z_A) \in A} \{z_A\}$,则 $\min\limits_{(x_A,y_A,z_A) \in A} \{z_A\} \leqslant \min\limits_{(x_B,y_B,z_B) \in B} \{z_B\}$,$\max\limits_{(x_B,y_B,z_B) \in B} \{z_B\} \leqslant \max\limits_{(x_A,y_A,z_A) \in A} \{z_A\}$,再由式(3.52)得 $\min\limits_{(x_A,y_A,z_A) \in A} \{z_A\} < z_C \leqslant \max\limits_{(x_A,y_A,z_A) \in A} \{z_A\}$。

(3) 如果 $j = \text{down}_A$,即 $z_B < \min\limits_{(x_A,y_A,z_A) \in A} \{z_A\}$,则 $\max\limits_{(x_B,y_B,z_B) \in B} \{z_B\} < \min\limits_{(x_A,y_A,z_A) \in A} \{z_A\}$,再由式(3.52)得 $z_C < \min\limits_{(x_A,y_A,z_A) \in A} \{z_A\}$。

根据以上 3 种情况得

第 3 章　方向关系描述及其定性推理

$$O_{i_1,j} \wedge O_{i_2,\text{between}} \rightarrow \begin{cases} O_{k_1,\text{up}_A} \vee O_{k_2,\text{up}_A} \vee \cdots \vee O_{k_l,\text{up}_A} & j = \text{up}_A \\ O_{k_1,\text{between}_A} \vee O_{k_2,\text{between}_A} \vee \cdots \vee O_{k_l,\text{between}_A} & j = \text{between}_A \\ O_{k_1,\text{down}_A} \vee O_{k_2,\text{down}_A} \vee \cdots \vee O_{k_l,\text{down}_A} & j = \text{down}_A \end{cases} \quad (3.53)$$

即 $O_{i_1,j} \wedge O_{i_2,\text{between}} \rightarrow O_{k_1,j} \vee O_{k_2,j} \vee \cdots \vee O_{k_l,j}$。

定向推理 $O_{i_1,\text{up}} \wedge O_{i_2,\text{between}}$ 的结果见表 3.14，$O_{i_1,\text{between}} \wedge O_{i_2,\text{between}}$ 的推理结果见表 3.12，$O_{i_1,\text{down}} \wedge O_{i_2,\text{between}}$ 的推理结果见表 3.15。

表 3.14　$O_{i_1,\text{up}} \wedge O_{i_2,\text{between}}$ 推理组合

表 3.15　$O_{i_1,\text{down}} \wedge O_{i_2,\text{between}}$ 推理组合

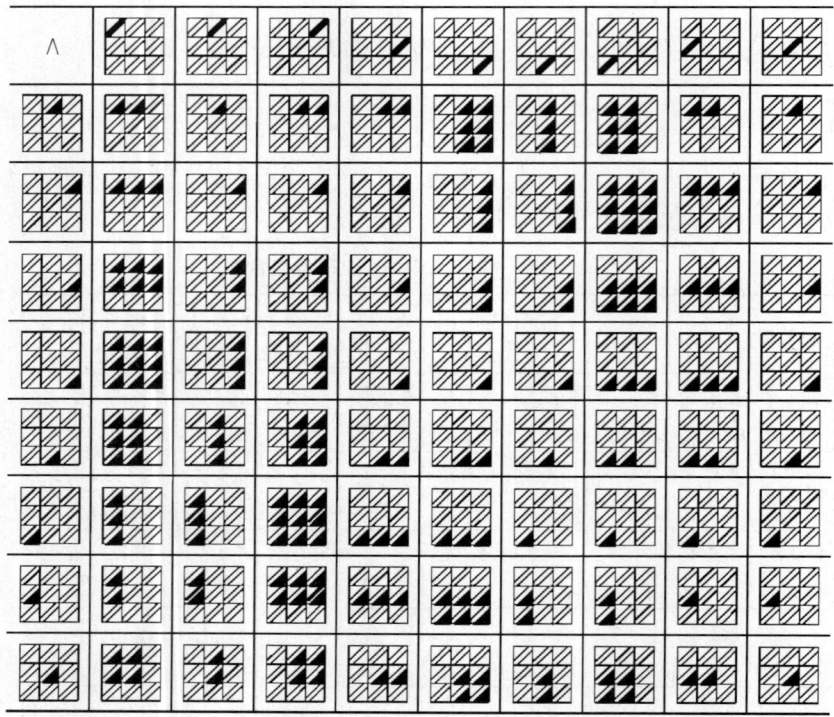

——式(3.46)的证明如下。

$O_{i_1,\text{between}} \wedge O_{i_2,j} \to \{i_1 \in DIRXY_A, \min\limits_{(x_A,y_A,z_A)\in A}\{z_A\} \leqslant z_B \leqslant \max\limits_{(x_A,y_A,z_A)\in A}\{z_A\}\} \cap$

$\{i_2 \in DIRXY_B, j \in DIRZ_B\}$ (3.54)

(1) 如果 $j=\text{up}_B$,即 $z_C > \max\limits_{(x_B,y_B,z_B)\in B}\{z_B\}$,再由式(3.54)得 $\min\limits_{(x_A,y_A,z_A)\in A}\{z_A\} \leqslant z_C \leqslant \max\limits_{(x_A,y_A,z_A)\in A}\{z_A\}$ 或 $z_C > \max\limits_{(x_A,y_A,z_A)\in A}\{z_A\}$。

(2) 如果 $j=\text{between}_B$,即 $\min\limits_{(x_B,y_B,z_B)\in B}\{z_B\} \leqslant z_C \leqslant \max\limits_{(x_B,y_B,z_B)\in B}\{z_B\}$,再由式(3.54)得 $\min\limits_{(x_A,y_A,z_A)\in A}\{z_A\} \leqslant z_C \leqslant \max\limits_{(x_A,y_A,z_A)\in A}\{z_A\}$。

(3) 如果 $j=\text{down}_B$,即 $z_C < \min\limits_{(x_B,y_B,z_B)\in B}\{z_B\}$,再由式(3.54)得 $z_C < \min\limits_{(x_A,y_A,z_A)\in A}\{z_A\}$ 或 $\min\limits_{(x_A,y_A,z_A)\in A}\{z_A\} \leqslant z_C \leqslant \max\limits_{(x_A,y_A,z_A)\in A}\{z_A\}$。

由以上 3 种情况得

$O_{i_1,\text{between}} \wedge O_{i_2,j} \to \begin{cases} O_{k_1,\text{up}} \vee \cdots \vee O_{k_l,\text{up}} \vee O_{k_1,\text{between}} \vee \cdots \vee O_{k_l,\text{between}} & j=\text{up}_B \\ O_{k_1,\text{between}} \vee \cdots \vee O_{k_l,\text{between}} & j=\text{between}_B \\ O_{k_1,\text{between}} \vee \cdots \vee O_{k_l,\text{between}} \vee O_{k_1,\text{down}} \vee \cdots \vee O_{k_l,\text{down}} & j=\text{down}_B \end{cases}$

(3.55)

第3章 方向关系描述及其定性推理

即 $O_{i_1,\text{between}} \wedge O_{i_2,j} \to O_{k_1,j} \vee O_{k_2,j} \vee \cdots \vee O_{k_l,j} \vee O_{k_1,\text{between}} \vee O_{k_2,\text{between}} \vee \cdots \vee O_{k_l,\text{between}}$ 。

$O_{i_1,\text{between}} \wedge O_{i_2,\text{up}}$ 推理结果见表 3.16，$O_{i_1,\text{between}} \wedge O_{i_2,\text{between}}$ 推理结果见表 3.12，$O_{i_1,\text{between}} \wedge O_{i_2,\text{down}}$ 推理结果见表 3.17。

表 3.16 $O_{i_1,\text{between}} \wedge O_{i_2,\text{up}}$ **推理组合**

表 3.17 $O_{i_1,\text{between}} \wedge O_{i_2,\text{down}}$ **推理组合**

续表

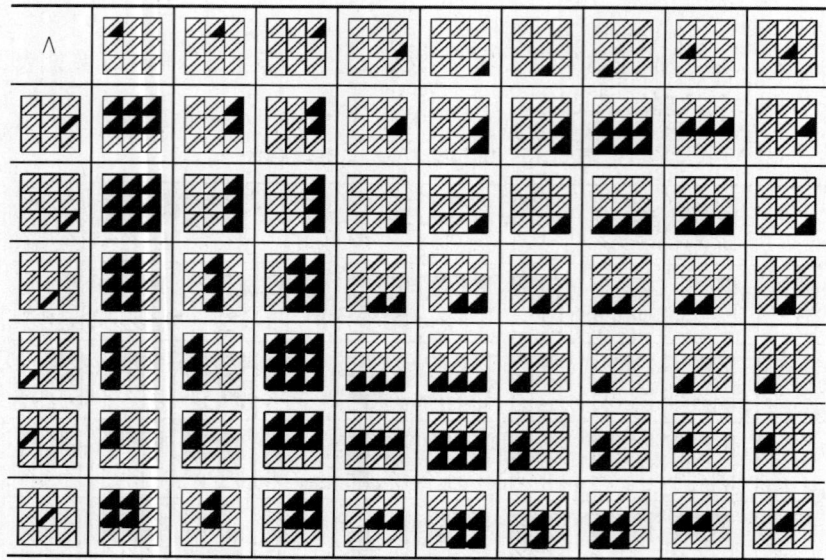

——式(3.47)的证明如下。

$O_{i_1,up} \wedge O_{i_2,down} = \{(x,y,z) | i_1 \in DIRXY_A, z_B > \max\{z_A | (x_A, y_A, z_A) \in A\}\} \cap$
$\quad \{(x_C, y_C, z_C) | i_2 \in DIRXY_B, z_C < \min\{z_B | (x_B, y_B, z_B) \in B\}\}$
$\rightarrow \{(x_C, y_C, z_C) | k \in DIRXY_A, z_C > \max\{z_A | (x_A, y_A, z_A) \in A\}$ or
$\quad \min\{z_A | (x_A, y_A, z_A) \in A\} \leqslant z_C \leqslant \max\{z_A | (x_A, y_A, z_A) \in A\}$ or
$\quad z_C < \min\{z_A | (x_A, y_A, z_A) \in A\}\}$
$\rightarrow O_{k_1,up} \vee \cdots \vee O_{k_l,up} \vee O_{k_1,between} \vee \cdots \vee O_{k_l,between} \vee O_{i_1,down} \vee \cdots \vee O_{i_l,down}$

(3.56)

$O_{i_1,down} \wedge O_{i_2,up} = \{(x_B, y_B, z_B) | i_1 \in DIRXY_A, z_B < \min_{(x_A, y_A, z_A) \in A}\{z_A\}\} \cap$
$\quad (x_C, y_C, z_C) | i_2 \in DIRXY_B, z_C > \max_{(x_B, y_B, z_B) \in B}\{z_B\}\}$
$\rightarrow \{(x_C, y_C, z_C) | k \in DIRXY_A, z_C < \min_{(x_A, y_A, z_A) \in A}\{x_A\}$ or
$\quad \min_{(x_A, y_A, z_A) \in A}\{x_A\} \leqslant z_C \leqslant \max_{(x_A, y_A, z_A) \in A}\{x_A\}$ or $z_C > \max_{(x_A, y_A, z_A) \in A}\{x_A\}\}$
$\rightarrow O_{k_1,up} \vee \cdots \vee O_{k_l,up} \vee O_{k_1,between} \vee \cdots \vee O_{k_l,between} \vee O_{k_1,down} \vee \cdots \vee O_{k_l,down}$

(3.57)

由式(3.56)和式(3.57)得 $O_{i_1,up} \wedge O_{i_2,down} = O_{i_1,down} \wedge O_{i_2,up}$。$O_{i_1,up} \wedge O_{i_2,down}$ 具体推理结果见表3.18,$O_{i_1,down} \wedge O_{i_2,up}$ 具体推理结果见表3.19。

式(3.44)说明:Z 方向关系满足传递性。式(3.45)说明:当 $dir(B,C)$ 的 Z 方向关系为 between 时,推理结果为 $dir(A,B)$ 的 Z 方向关系。式(3.46)说明:当

第3章 方向关系描述及其定性推理　　79

$dir(A,B)$ 的 Z 方向关系为 between 时，定性推理的结果为 $dir(A,B)$ 与 $dir(B,C)$ 的 Z 方向关系之和。式(3.47)说明：当 $dir(A,B)$ 与 $dir(B,C)$ 的 Z 方向关系是互为相反的方向关系时，定性推理的结果为 up, between 和 down, 没有提供 Z 方向关系。

表 3.18　$O_{i_1,\text{up}} \wedge O_{i_2,\text{down}}$ 推理组合

表 3.19　$O_{i_1,\text{down}} \wedge O_{i_2,\text{up}}$ 推理组合

续表

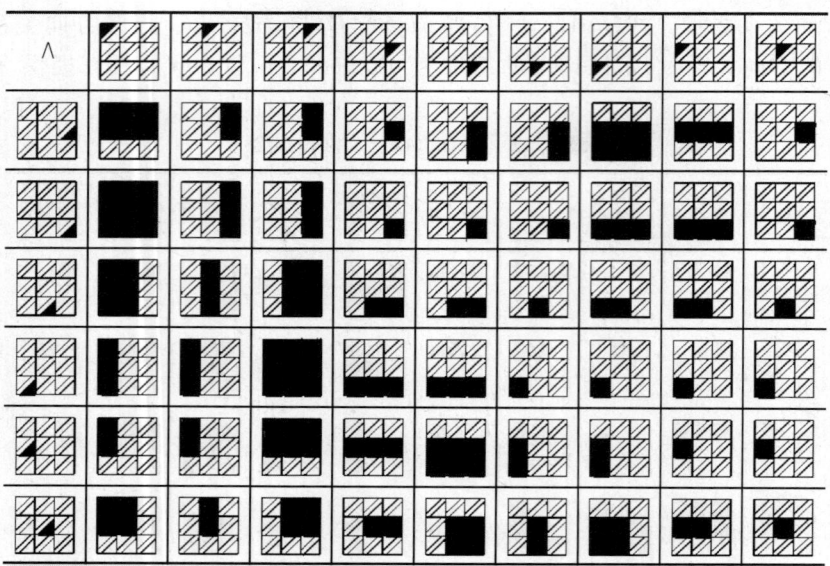

由复合 X-Y 方向关系推理的规律和 Z 方向关系推理规律,可得如下规律。

式(3.44)分别与式(3.27)至式(3.41)复合,得到式(3.58)至式(3.72)。$dir(A,C)$ 的 Z 方向关系规律如式(3.44)所示,X-Y 方向关系规律分别如式(3.27)至式(3.41)所示。其中 $j \in \{\text{up}, \text{between}, \text{down}\}$。

$$O_{i,j} \wedge O_{i,j} \rightarrow O_{i,j} \tag{3.58}$$

$$O_{i,j} \wedge O_{\text{same},j} \rightarrow O_{i,j} \tag{3.59}$$

$$O_{i,j} \wedge O_{N,j} \rightarrow O_{i,j} \tag{3.60}$$

式中,$i \in \{\text{NW}, \text{N}, \text{NE}\}$。

$$O_{W,j} \wedge O_{N,j} \rightarrow O_{W,j} \vee O_{NW,j} \tag{3.61}$$

$$O_{\text{same},j} \wedge O_{N,j} \rightarrow O_{\text{same},j} \vee O_{N,j} \tag{3.62}$$

$$O_{E,j} \wedge O_{N,j} \rightarrow O_{E,j} \vee O_{NE,j} \tag{3.63}$$

$$O_{SW,j} \wedge O_{N,j} \rightarrow O_{SW,j} \vee O_{W,j} \vee O_{NW,j} \tag{3.64}$$

$$O_{S,j} \wedge O_{N,j} \rightarrow O_{S,j} \vee O_{\text{same},j} \vee O_{N,j} \tag{3.65}$$

$$O_{SE,j} \wedge O_{N,j} \rightarrow O_{SE,j} \vee O_{E,j} \vee O_{NE,j} \tag{3.66}$$

$$O_{N,j} \wedge O_{NE,j} \rightarrow O_{N,j} \vee O_{NE,j} \tag{3.67}$$

$$O_{E,j} \wedge O_{NE,j} \rightarrow O_{E,j} \vee O_{NE,j} \tag{3.68}$$

$$O_{NW,j} \wedge O_{NE,j} \rightarrow O_{NW,j} \vee O_{N,j} \vee O_{NE,j} \tag{3.69}$$

$$O_{SE,j} \wedge O_{NE,j} \rightarrow O_{SE,j} \vee O_{E,j} \vee O_{NE,j} \tag{3.70}$$

$$O_{\text{same},j} \wedge O_{NE,j} \rightarrow O_{\text{same},j} \vee O_{NE,j} \vee O_{N,j} \vee O_{E,j} \tag{3.71}$$

$$O_{SW,j} \wedge O_{NE,j} \to O_{N,j} \vee O_{NE,j} \vee O_{E,j} \vee O_{SE,j} \vee O_{S,j} \vee O_{SW,j} \vee$$
$$O_{W,j} \vee O_{NW,j} \vee O_{same,j} \tag{3.72}$$

式(3.58)说明单方向关系推理满足传递性。在三维空间中,方向关系是由 X-Y 方向关系和 Z 方向关系复合而成。根据式(3.44)可知,Z 方向关系具有传递性;根据式(3.27)可知,X-Y 方向关系具有传递性。因此,当 $dir(A,B) = dir(B,C)$ 时,$dir(A,B) \wedge dir(A,B) \to dir(A,B)$。

式(3.45)分别与式(3.27)至式(3.41)复合时,得到式(3.73)至式(3.87),$dir(A,C)$ 的 Z 方向关系规律如式(3.45)所示,X-Y 方向关系规律分别如式(3.27)至式(3.41)所示。其中 $j \in \{up, between, down\}$。

$$O_{i,j} \wedge O_{i,between} \to O_{i,j} \tag{3.73}$$
$$O_{i,j} \wedge O_{same,between} \to O_{i,j} \tag{3.74}$$
$$O_{i,j} \wedge O_{N,between} \to O_{i,j} \tag{3.75}$$

式中,$i \in \{NW, N, NE\}$。

$$O_{W,j} \wedge O_{N,between} \to O_{W,j} \vee O_{NW,j} \tag{3.76}$$
$$O_{same,j} \wedge O_{N,between} \to O_{same,j} \vee O_{N,j} \tag{3.77}$$
$$O_{E,j} \wedge O_{N,between} \to O_{E,j} \vee O_{NE,j} \tag{3.78}$$
$$O_{SW,j} \wedge O_{N,between} \to O_{SW,j} \vee O_{W,j} \vee O_{NW,j} \tag{3.79}$$
$$O_{S,j} \wedge O_{N,between} \to O_{S,j} \vee O_{same,j} \vee O_{N,j} \tag{3.80}$$
$$O_{SE,j} \wedge O_{N,between} \to O_{SE,j} \vee O_{E,j} \vee O_{NE,j} \tag{3.81}$$
$$O_{N,j} \wedge O_{NE,between} \to O_{N,j} \vee O_{NE,j} \tag{3.82}$$
$$O_{E,j} \wedge O_{NE,between} \to O_{E,j} \vee O_{NE,j} \tag{3.83}$$
$$O_{NW,j} \wedge O_{NE,between} \to O_{NW,j} \vee O_{N,j} \vee O_{NE,j} \tag{3.84}$$
$$O_{SE,j} \wedge O_{NE,between} \to O_{SE,j} \vee O_{E,j} \vee O_{NE,j} \tag{3.85}$$
$$O_{same,j} \wedge O_{NE,between} \to O_{same,j} \vee O_{NE,j} \vee O_{N,j} \vee O_{E,j} \tag{3.86}$$
$$O_{SW,j} \wedge O_{NE,between} \to O_{N,j} \vee O_{NE,j} \vee O_{E,j} \vee O_{SE,j} \vee$$
$$O_{S,j} \vee O_{SW,j} \vee O_{W,j} \vee O_{NW,j} \vee O_{same,j} \tag{3.87}$$

式(3.46)分别与式(3.27)至式(3.41)复合,得到式(3.88)至式(3.102)。$dir(A,C)$ 的 Z 方向关系如式(3.46)所示,X-Y 方向关系规律分别如式(3.27)至式(3.41)所示。其中 $j \in \{up, between, down\}$。

$$O_{i,between} \wedge O_{i,j} \to O_{i,j} \vee O_{i,between} \tag{3.88}$$
$$O_{i,between} \wedge O_{same,j} \to O_{i,j} \vee O_{i,between} \tag{3.89}$$
$$O_{i,between} \wedge O_{N,j} \to O_{i,j} \vee O_{i,between} \tag{3.90}$$

式中,$i \in \{NW, N, NE\}$。

$$O_{W,between} \wedge O_{N,j} \to O_{W,j} \vee O_{W,between} \vee O_{NW,j} \vee O_{NW,between} \tag{3.91}$$

$$O_{\text{same,between}} \wedge O_{\text{N},j} \rightarrow O_{\text{same},j} \vee O_{\text{same,between}} \vee O_{\text{N},j} \vee O_{\text{N,between}} \tag{3.92}$$

$$O_{\text{E,between}} \wedge O_{\text{N},j} \rightarrow O_{\text{E},j} \vee O_{\text{E,between}} \vee O_{\text{NE},j} \vee O_{\text{NE,between}} \tag{3.93}$$

$$O_{\text{SW,between}} \wedge O_{\text{N},j} \rightarrow O_{\text{SW},j} \vee O_{\text{SW,between}} \vee O_{\text{W},j} \vee O_{\text{W,between}} \vee$$
$$O_{\text{NW},j} \vee O_{\text{NW,between}} \tag{3.94}$$

$$O_{\text{S,between}} \wedge O_{\text{N},j} \rightarrow O_{\text{S},j} \vee O_{\text{S,between}} \vee O_{\text{same},j} \vee O_{\text{same,between}} \vee$$
$$O_{\text{N},j} \vee O_{\text{N,between}} \tag{3.95}$$

$$O_{\text{SE,between}} \wedge O_{\text{N},j} \rightarrow O_{\text{SE},j} \vee O_{\text{SE,between}} \vee O_{\text{E},j} \vee O_{\text{E,between}} \vee$$
$$O_{\text{NE},j} \vee O_{\text{NE,between}} \tag{3.96}$$

$$O_{\text{N,between}} \wedge O_{\text{NE},j} \rightarrow O_{\text{N},j} \vee O_{\text{N,between}} \vee O_{\text{NE},j} \vee O_{\text{NE,between}} \tag{3.97}$$

$$O_{\text{E,between}} \wedge O_{\text{NE},j} \rightarrow O_{\text{E},j} \vee O_{\text{E,between}} \vee O_{\text{NE},j} \vee O_{\text{NE,between}} \tag{3.98}$$

$$O_{\text{NW,between}} \wedge O_{\text{NE},j} \rightarrow O_{\text{NW},j} \vee O_{\text{NW,between}} \vee O_{\text{N},j} \vee O_{\text{N,between}} \vee$$
$$O_{\text{NE},j} \vee O_{\text{NE,between}} \tag{3.99}$$

$$O_{\text{SE,between}} \wedge O_{\text{NE},j} \rightarrow O_{\text{SE},j} \vee O_{\text{SE,between}} \vee O_{\text{E},j} \vee O_{\text{E,between}} \vee$$
$$O_{\text{NE},j} \vee O_{\text{NE,between}} \tag{3.100}$$

$$O_{\text{same,between}} \wedge O_{\text{NE},j} \rightarrow O_{\text{same},j} \vee O_{\text{same,between}} \vee O_{\text{NE},j} \vee O_{\text{NE,between}} \vee O_{\text{N},j} \vee$$
$$O_{\text{N,between}} \vee O_{\text{E},j} \vee O_{\text{E,between}} \tag{3.101}$$

$$O_{\text{SW,between}} \wedge O_{\text{NE},j} \rightarrow O_{\text{N},j} \vee O_{\text{NE},j} \vee O_{\text{E},j} \vee O_{\text{SE},j} \vee O_{\text{S},j} \vee O_{\text{SW},j} \vee O_{\text{W},j} \vee$$
$$O_{\text{NW},j} \vee O_{\text{same},j} \vee O_{\text{N,between}} \vee O_{\text{NE,between}} \vee O_{\text{E,between}} \vee$$
$$O_{\text{SE,between}} \vee O_{\text{S,between}} \vee O_{\text{SW,between}} \vee O_{\text{W,between}} \vee$$
$$O_{\text{NW,between}} \vee O_{\text{same,between}} \tag{3.102}$$

当式(3.47)分别与式(3.27)至式(3.41)复合时,由于式(3.47)的推理结果没有提供 Z 方向关系约束,复合后的规律是 X-Y 方向关系推理规律。如

$$O_{\text{N,up}} \wedge O_{\text{NE,down}} \rightarrow O_{\text{N,up}} \vee O_{\text{N,between}} \vee O_{\text{N,down}} \vee O_{\text{NE,up}} \vee O_{\text{NE,between}} \vee O_{\text{NE,down}}$$

$$O_{\text{SW,up}} \wedge O_{\text{N,down}} \rightarrow O_{\text{SW,up}} \vee O_{\text{SW,between}} \vee O_{\text{SW,down}} \vee O_{\text{W,up}} \vee O_{\text{W,between}} \vee O_{\text{W,down}} \vee O_{\text{NW,up}} \vee$$
$$O_{\text{NW,between}} \vee O_{\text{NW,down}}$$

式(3.41)推理的结果没有 X-Y 方向关系约束,当与式(3.44)至式(3.46)复合时,结果只有 Z 方向关系约束,如式(3.72)、式(3.87)和式(3.102)。当式(3.41)和式(3.47)复合时,推理结果没有约束,27 个方向都有可能。

第4章 定性距离描述及其推理

距离是人们在日常生活中经常使用的概念,它描述了两个实体之间的远近或亲疏程度。获取距离的方法主要有:根据已知点的坐标计算获得,通过实际测量获得,通过目测获得或者是通过想象获得。称通过计算或实际测量获得的距离为实际距离或客观距离(Briggs,1973a)。在大比例尺空间中,人们对相距很远以致相互之间无法看到的两地之间距离的认识称为认知距离,在估计两地距离的过程中可以看到两地的距离估计称为感觉距离(Montello,1991)。感觉距离为认知距离的估计提供了重要的信息(Hong,1994)。认知距离受所使用的推理机制以及个人经验等多个因素的影响(Briggs,1973b;Lunberg,1973;Maki,1981;Sadalla et al,1980a,1980b;Thorndyke,1981;Kirasic et al,1984;McNamara et al,1984;McNamara et al,1989;Sergent,1991),并且认知距离不遵循欧氏距离的3个基本特征(对称性、自反性和三角形不等式)(Cadwallader,1979;Garling et al,1991)。由于认知距离与实际距离常常有一定的偏差,在认知空间中不能保持全局一致性,因此对现实世界的解释可能是正确的,也可能是扭曲的,有时甚至是相互矛盾的。由于无法准确地确定认知距离的大小,因此应该考虑认知距离的不确定性,在地理空间中将认知距离对应为度量值的一个区间。

目前,二维空间中定性距离划分空间的方法是根据目标对象与参照物的距离划分参照物所在的空间成几个部分,如,2个部分:近、远;3个部分:近、一般、远;4个部分:很近、近、远、很远。通常定性距离满足3个约束条件:①距离范围单调递增;②距离范围大于前面所有距离范围之和;③吸收率。根据定性距离的3个约束条件研究同方向定性距离推理和反方向定性距离推理。

本章的研究重点是在三维空间中建立定性距离描述系统,并根据定性距离的3个约束条件研究定性距离推理。在研究定性距离推理时,引入区间数的概念,通过对区间数的代数运算研究同方向定性距离推理和反方向定性距离推理。

§4.1 定性距离描述

4.1.1 距离

在地理空间中,距离反映了空间物体间的几何接近程度。数学上严格的距离定义为:

定义 4.1 （刘炳初,2001）设 X 是任一非空集,对 X 中任意两点 x、y 有一个实数 $d(x,y)$ 与之对应,且满足

(1) $d(x,y) \geqslant 0$,且 $d(x,y)=0$,当且仅当 $x=y$（自反性）；

(2) $d(x,y)=d(y,x)$（对称性）；

(3) $d(x,y) \leqslant d(x,z)+d(z,y)$（三角形不等式）；

称 $d(x,y)$ 为 X 中的一个距离。

当 $X=\mathbf{R} \times \mathbf{R}$ 时,则 $d(x,y)$ 为二维空间中的距离。当 $X=\mathbf{R} \times \mathbf{R} \times \mathbf{R}$ 时,$d(x,y)$ 为三维空间中的距离。常用的距离有欧氏距离、Minkowski p-metrics 距离和 Manhattan 距离等（Gatrell，1983）。

在二维空间中,计算点 $A(x_A, y_A)$ 和 $B(x_B, y_B)$ 之间距离的常用方法有：

——欧氏距离,$d(A,B)=\sqrt{(x_A-x_B)^2+(y_A-y_B)^2}$。在二维笛卡儿系统下,欧氏距离对应于两物体间的最短距离。

——Minkowski p-metrics 距离,$d(A,B)=((x_A-x_B)^p+(y_A-y_B)^p)^{1/p}$。

——Manhattan 距离,$d(A,B)=|x_A-x_B|+|y_A-y_B|$。

在三维空间中,计算点 $A(x_A, y_A, z_A)$ 和点 $B(x_B, y_B, z_B)$ 之间距离的常用方法有：

——欧氏距离,$d(A,B)=((x_A-x_B)^2+(y_A-y_B)^2+(z_A-z_B)^2)^{1/2}$。

——Minkowski p-metrics 距离,$d(A,B)=((x_A-x_B)^p+(y_A-y_B)^p+(z_A-z_B)^p)^{1/p}$。

——Manhattan 距离,$d(A,B)=|x_A-x_B|+|y_A-y_B|+|z_A-z_B|$。

地理空间中的距离所描述的对象一定是位于地理空间中,也就是说它具有空间概念,是基于地理位置的,反映了空间物体间的几何接近程度。在地理空间中,不仅要计算点状物体间的距离,还要计算非点状物体间的距离。计算非点状物体间距离的方法主要有：两物体重心间的距离,即两物体间的平均距离（Koshizuka et al，1991）；两物体间的最短距离（Peuquet，1992）；两物体间的最远距离等。

定义 4.1 是根据已知点的坐标计算距离,称这种通过测量或在某一参照系下根据两个物体的已知坐标进行计算获得的定量值为定量距离（Santos et al，2005）,定量距离是用数值描述的距离,又称为度量距离。度量距离通常以人们设定的度量系统（如 m 或 km）为基础。度量系统中的度量单位是精确的,不会引起任何误解。人们常说的"5 步远"、"步行 10 分钟的路程"等是定量距离的范围,而不是单个距离值。定量距离描述的一个优势是当仔细测量时,定量距离精确度高。但是,当无法仔细测量时,定量距离的准确度低,如"大约 5 米",这样用定量距离的一个区间代替了定量距离,称为定性距离。人们用这种定性距离进行交流基本没有困难,这可能是在日常的生活中这种描述方法已经满足人们交流的需求（Hong et al，1995）。

4.1.2 定性距离

在地理空间中,设参照物为 A,目标对象为 B,目标对象 B 与参照物 A 间的距离记为 $d(A,B)$。根据 $d(A,B)$ 的大小,将 $d(A,B)$ 进行划分,如图 4.1 所示,$d(A,B)$ 可划分为 $(a_0,a_1],(a_1,a_2],(a_2,a_3],(a_3,a_4],(a_4,a_5],\cdots$用自然语言将上述距离区间分别描述为 $q_0,q_1,q_2,q_3,q_4,\cdots$。

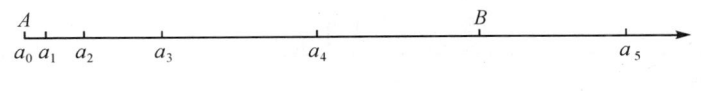

图 4.1 距离区间划分

定义 4.2 称描述距离区间 $(a_i,a_{i+1}]$ 的自然语言 q_i 为定性距离 q_i。

当 $d(A,B)\in(a_i,a_{i+1}],i\in\{0,1,2,\cdots\}$时,称目标对象 B 与参照物 A 间的定性距离为 q_i,用符号表示为 $d_{AB}=q_i$。

定性距离是用自然语言(如远、近等)描述的命题(Hong,1994)。"远"描述较远的距离,"近"描述较近的距离,但它们没有传递出具体的度量信息。人们常用"很近、近、远、很远"等描述从最近到最远的距离,这些术语有一定的顺序性,"很近"描述较短的定量距离,"很远"描述较远的定量距离。与定量距离相比,定性距离缺乏表达力。由于定性距离用较少的术语描述定量距离相应的范围,因此,定性距离对应于定量距离的一个集合,相同的定性距离可以描述不同的定量距离集合,而且一个定量距离也许用不同的定性距离描述。例如,两个地区相距 10 km,对于步行的人来说也许"很远",对骑自行车的人来说"近",而对于开车的人来说"很近"。从理论上讲定性距离的个数是没有限制的,但实际上用自然语言描述的定性距离的个数是相当有限的,人们能够区分的定性距离的个数约为 7 ± 2。

定性距离包含 3 个元素,即目标对象 B、参照对象 A 和参照框架(Hernández et al,1995)。定性距离可用谓词集 $\{\leqslant,=,\geqslant\}$ 比较。根据不同的区分程度,可以将参照物所在的空间分为 2 个部分:近、远;3 个部分:近、中间和远;4 个部分:很近、近、远和很远;5 个部分:很近、近、中间、远和很远。下面以 5 个划分为例说明定性距离。

令 $\delta_i=a_{i+1}-a_i,i\in\{0,1,2,3,4\}$表示区间 $(a_i,a_{i+1}]$的长度。在确定距离区间时,必须满足下列条件:相邻距离区间的长度之比为一个常数 k,即 $k=\delta_i/\delta_{i-1}$。这样以同样的比例放大或缩小距离区间,定性距离合成的结果具有鲁棒性(Hong,1994)。因此,同一个区间长度比对应多个不同的距离区间,如 $k=2$ 所对应的距离区间可以是 $(0,1],(1,3],(3,7],(7,15],(15,31],\cdots$也可以是 $(0,2],(2,6],(6,$

14],(14,30],(30,62],…还可以是(0,100],(100,300],(300,700],(700,1 500],(1 500,3 100],…后两个分别将距离区间的长度放大 2 倍和 100 倍,但相邻距离区间之间的长度之比都是 $k=2$。为规范定性距离的距离区间,称第一个距离区间为 (0,1] 的定性距离区间为标准定性距离区间,即称 (0,1],(1,3],(3,7],(7,15],(15,31],…为 $k=2$ 时的标准距离区间,表 4.1 给出了不同的距离区间长度比所对应的标准距离区间。

表 4.1 标准距离区间表

k	q_0	q_1	q_2	q_3	q_4
1	(0,1]	(1,2]	(2,3]	(3,4]	(4,5]
2	(0,1]	(1,3]	(3,7]	(7,15]	(15,31]
3	(0,1]	(1,4]	(4,13]	(13,40]	(40,121]
4	(0,1]	(1,5]	(5,21]	(21,85]	(85,341]
5	(0,1]	(1,6]	(6,31]	(31,156]	(156,781]
6	(0,1]	(1,7]	(7,43]	(43,259]	(259,1 555]
7	(0,1]	(1,8]	(8,57]	(57,400]	(400,2 801]
8	(0,1]	(1,9]	(9,73]	(73,585]	(585,4 681]
9	(0,1]	(1,10]	(10,91]	(91,820]	(820,7 381]
10	(0,1]	(1,11]	(11,111]	(111,1 111]	(1 111,11 111]
20	(0,1]	(1,21]	(21,421]	(421,8 421]	(8 421,168 421]
50	(0,1]	(1,51]	(51,2 551]	(2 551,127 551]	(127 551,6 377 551]
100	(0,1]	(1,101]	(101,10 101]	(10 101,1 010 101]	(1 010 101,101 010 101]

§4.2 定性距离间的定性推理

4.2.1 定性推理

定性距离的推理是根据参照物 A 与目标对象 B 间的定性距离,以及参照物 B 与目标对象 C 间的定性距离,推导 A 与目标对象 C 间的定性距离,用公式表示为

$$d_{AB} \wedge d_{BC} \to d_{AC} \tag{4.1}$$

令 $\Delta_i = \delta_0 + \cdots + \delta_i, i > 0$,如图 4.2 所示(Hernández,1995)。那么,定性距离应满足下面 3 个约束条件:

(1) 距离范围单调递增,即 $\delta_0 \leqslant \delta_1 \leqslant \delta_2 \leqslant \delta_3 \leqslant \cdots \leqslant \delta_n$。

(2) 范围限制,即 $\delta_i \geqslant \Delta_{i-1}$。

(3) 吸收律,即如果 $\delta_i \gg \delta_j$,则 $\delta_i \pm \delta_j \approx \delta_i$。

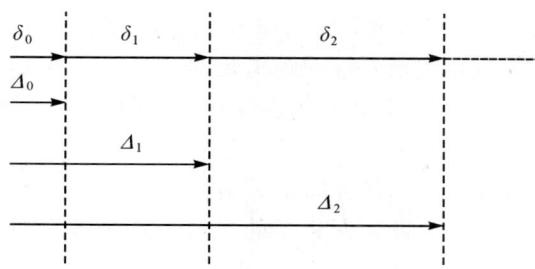

图 4.2 Δ_i 的定义

根据距离范围的单调性,可得

$$k = \frac{\delta_i}{\delta_{i-1}} \geqslant 1 \qquad (4.2)$$

各定性距离的距离范围分别为:$\delta_1 = k\delta_0, \delta_2 = k^2\delta_0, \delta_3 = k^3\delta_0, \delta_4 = k^4\delta_0$。对于不同的 k,具体的定性距离的距离范围见表 4.2。

表 4.2 距离范围表

k	δ_0	$\delta_1 = k\delta_0$	$\delta_2 = k^2\delta_0$	$\delta_3 = k^3\delta_0$	$\delta_4 = k^4\delta_0$
$k=1$	δ_0	$\delta_1 = \delta_0$	$\delta_2 = \delta_0$	$\delta_3 = \delta_0$	$\delta_4 = \delta_0$
$k=2$	δ_0	$\delta_1 = 2\delta_0$	$\delta_2 = 4\delta_0$	$\delta_3 = 8\delta_0$	$\delta_4 = 16\delta_0$
$k=3$	δ_0	$\delta_1 = 3\delta_0$	$\delta_2 = 9\delta_0$	$\delta_3 = 27\delta_0$	$\delta_4 = 81\delta_0$
$k=4$	δ_0	$\delta_1 = 4\delta_0$	$\delta_2 = 16\delta_0$	$\delta_3 = 64\delta_0$	$\delta_4 = 256\delta_0$
$k=5$	δ_0	$\delta_1 = 5\delta_0$	$\delta_2 = 25\delta_0$	$\delta_3 = 125\delta_0$	$\delta_4 = 625\delta_0$
$k=6$	δ_0	$\delta_1 = 6\delta_0$	$\delta_2 = 36\delta_0$	$\delta_3 = 216\delta_0$	$\delta_4 = 1\,296\delta_0$
$k=7$	δ_0	$\delta_1 = 7\delta_0$	$\delta_2 = 49\delta_0$	$\delta_3 = 343\delta_0$	$\delta_4 = 2\,401\delta_0$
$k=8$	δ_0	$\delta_1 = 8\delta_0$	$\delta_2 = 64\delta_0$	$\delta_3 = 512\delta_0$	$\delta_4 = 4\,096\delta_0$
$k=9$	δ_0	$\delta_1 = 9\delta_0$	$\delta_2 = 81\delta_0$	$\delta_3 = 729\delta_0$	$\delta_4 = 6\,561\delta_0$
$k=10$	δ_0	$\delta_1 = 10\delta_0$	$\delta_2 = 100\delta_0$	$\delta_3 = 1\,000\delta_0$	$\delta_4 = 10\,000\delta_0$
$k=20$	δ_0	$\delta_1 = 20\delta_0$	$\delta_2 = 400\delta_0$	$\delta_3 = 8\,000\delta_0$	$\delta_4 = 160\,000\delta_0$
$k=50$	δ_0	$\delta_1 = 50\delta_0$	$\delta_2 = 2\,500\delta_0$	$\delta_3 = 125\,000\delta_0$	$\delta_4 = 6\,250\,000\delta_0$
$k=100$	δ_0	$\delta_1 = 100\delta_0$	$\delta_2 = 10\,000\delta_0$	$\delta_3 = 1\,000\,000\delta_0$	$\delta_4 = 100\,000\,000\delta_0$

根据图 4.2 可知,$\Delta_{i-1} = \delta_0 + \cdots + \delta_{i-1}, \delta_i = \delta_0 + (k-1)\Delta_{i-1}$,因此,当 $k \geqslant 2$ 时,约束条件(2)成立。根据约束条件(2),得

$$A_i = \sum_{k=0}^{i-1} \delta_k = \Delta_{i-1} \leqslant \delta_i \qquad (4.3)$$

在吸收律(3)中,应根据具体环境定义二元关系"$\gg$"。

定义 4.3 对于任意给定的正数 $\varepsilon(0 < \varepsilon < 1)$,如果满足 $\delta_j/\delta_i \leqslant \varepsilon$,则有 $\delta_i \gg \delta_j$。

注意:

(1)根据具体情况理解二元关系"$\gg$"的含义。如果令 $\varepsilon = 0.1$,则只要 $\delta_j/\delta_i \leqslant 0.1$,

就有 $\delta_i \pm \delta_j \approx \delta_i$；如果 $\varepsilon=0.01$，则只要 $\delta_j/\delta_i \leqslant 0.01$，就有 $\delta_i \pm \delta_j \approx \delta_i$。

(2)由于相邻的定性距离所对应的距离范围之比为常数 k，即 $k=\delta_{i+1}/\delta_i$ 是常数，因此，当 $k \geqslant 10$ 时，如果 $\varepsilon=0.1$，对任意的 $i,j \in \mathbf{N}, i \neq j$，均有 $\delta_i \pm \delta_j \approx \delta_i$；如果 $\varepsilon=0.01$，只要 $|i-j| \geqslant 2$，就有 $\delta_i \pm \delta_j \approx \delta_i$。

定义 4.4 在实数域 $\mathbf{R}$ 中，称区间 $\bar{a}=[a_l,a_r]$ 为区间数。
$\mathbf{R}$ 上的区间数，记为 $\bar{\mathbf{R}}$。设 $*$ 为 $\bar{\mathbf{R}}$ 上的二元算子，对于任意 $\bar{a},\bar{b} \in \bar{\mathbf{R}}$，有
$$\bar{a} * \bar{b} = \{z | z = x * y, x \in \bar{a}, y \in \bar{b}\} \tag{4.4}$$

因此
$$\bar{a}+\bar{b}=[a_l,a_r]+[b_l,b_r]=[a_l+b_l,a_r+b_r] \tag{4.5}$$
$$\bar{a}-\bar{b}=[a_l,a_r]-[b_l,b_r]=[a_l-b_r,a_r-b_l] \tag{4.6}$$
$$(a_i,a_{i+1}]+(a_j,a_{j+1}]=(a_i+a_j,a_{i+1}+a_{j+1}] \tag{4.7}$$
$$(a_i,a_{i+1}]-(a_j,a_{j+1}]=(a_i-a_{j+1},a_{i+1}-a_j) \tag{4.8}$$

4.2.2 同方向关系推理

设 $d_{AB}=q_i, d_{BC}=q_j$，当 q_i 与 q_j 方向相同时(图 4.3)，根据式(4.7)研究定性距离推理 $q_i \wedge q_j$(Hernández,1995)。

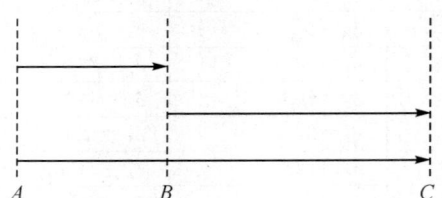

图 4.3 同方向定性距离推理

(1)当定性距离满足单调递增约束条件时，设 $i \leqslant j$，则
$$a_{i+1}+a_{j+1}=\sum_{l=0}^{i}\delta_l+\sum_{l=0}^{j}\delta_l \leqslant \sum_{l=j+1}^{i+j+1}\delta_l+\sum_{l=0}^{j}\delta_l=a_{j+i+2} \tag{4.9}$$

由式(4.7)和式(4.9)可知
$$(a_i,a_{i+1}]+(a_j,a_{j+1}] \subseteq (a_j,a_{j+i+2}] \tag{4.10}$$

由式(4.10)和定性距离的定义，可知
$$q_i \wedge a_j \rightarrow q_j \vee \cdots \vee q_{j+i+1} \tag{4.11}$$

当定性距离满足单调递增约束条件时，根据式(4.11)得
$$q_i \wedge a_j \rightarrow q_{\max(i,j)} \vee \cdots \vee q_{j+i+1} \tag{4.12}$$

当定性距离满足单调递增约束条件时，根据式(4.12)得到同方向定性距离组合推理表 4.3。

(2)当定性距离满足约束条件单调递增和范围限制时，设 $j \geqslant i$，则

第 4 章 定性距离描述及其推理

$a_{j+1} - a_j = \delta_j \geqslant \delta_i \geqslant a_i$,从而

$$a_i + a_j \in (a_j, a_{j+1}] \tag{4.13}$$

由式(4.13)可知

$$a_{i+1} + a_{j+1} \in (a_{j+1}, a_{j+2}] \tag{4.14}$$

由式(4.7),式(4.13)和式(4.14)得

$$(a_i, a_{i+1}] + (a_j, a_{j+1}] \subseteq (a_j, a_{j+2}] \tag{4.15}$$

根据定性距离的定义得

$$q_i \wedge q_j \rightarrow q_j \vee q_{j+1} \tag{4.16}$$

表 4.3 距离范围单调递增的同方向定性距离推理组合

$\wedge$	q_0	q_1	q_2	q_3	q_4
q_0	q_0, q_1	q_1, q_2	q_2, q_3	q_3, q_4	q_4
q_1	q_1, q_2	q_1, q_2, q_3	q_2, q_3, q_4	q_3, q_4	q_4
q_2	q_2, q_3	q_2, q_3, q_4	q_2, q_3, q_4	q_3, q_4	q_4
q_3	q_3, q_4	q_3, q_4	q_3, q_4	q_3, q_4	q_4
q_4	q_4	q_4	q_4	q_4	q_4

根据式(4.16),当距离范围单调递增,且满足 $\delta_i \geqslant \Delta_{i-1}(i>0)$ 时,两同方向的定性距离推理结果为

$$q_i \wedge q_j \rightarrow q_{\max\{i,j\}} \vee q_{\max\{i,j\}+1} \tag{4.17}$$

根据式(4.17)得到定性距离组合推理表 4.4。

表 4.4 距离范围单调递增,且满足范围限制的同方向定性距离推理组合

$\wedge$	q_0	q_1	q_2	q_3	q_4
q_0	q_0, q_1	q_1, q_2	q_2, q_3	q_3, q_4	q_4
q_1	q_1, q_2	q_1, q_2	q_2, q_3	q_3, q_4	q_4
q_2	q_2, q_3	q_2, q_3	q_2, q_3	q_3, q_4	q_4
q_3	q_3, q_4	q_3, q_4	q_3, q_4	q_3, q_4	q_4
q_4	q_4	q_4	q_4	q_4	q_4

根据吸收律,当 $\varepsilon = \frac{1}{k^2}$ 时,只要 $\frac{\delta_i}{\delta_j} \leqslant \varepsilon$,就有 $\delta_i + \delta_j \approx \delta_j$,即当 $\varepsilon = \frac{1}{k^2}$ 时,只要 $j - i \geqslant 2$,就有 $\delta_i + \delta_j \approx \delta_j$。所以

$$a_i + a_j = \sum_{k=0}^{i-1} \delta_k + \sum_{k=0}^{j-1} \delta_k \approx \sum_{k=0}^{j-1} \delta_k = a_j$$

同理,得 $a_{i+1} + a_{j+1} \approx a_{j+1}$。

由式(4.7)得

$$(a_i, a_{i+1}] + (a_j, a_{j+1}] \approx (a_j, a_{j+1}]$$

因此，当 $\varepsilon=\dfrac{1}{k^2}$ 时，只要 $j-i\geqslant 2$，就有

$$q_i \wedge q_j \to q_j \tag{4.18}$$

根据式(4.17)和式(4.18)，得到同方向定性距离组合推理表4.5。

表4.5 距离范围单调递增，且满足范围限制和吸收律的同方向定性距离推理组合

$\wedge$	q_0	q_1	q_2	q_3	q_4
q_0	q_0,q_1	q_1,q_2	q_2	q_3	q_4
q_1	q_1,q_2	q_1,q_2	q_2,q_3	q_3	q_4
q_2	q_2	q_2,q_3	q_2,q_3	q_3,q_4	q_4
q_3	q_3	q_3	q_3,q_4	q_3,q_4	q_4
q_4	q_4	q_4	q_4	q_4	q_4

4.2.3 反方向关系推理

当两个方向相反的定性距离合成时，根据式(4.8)研究定性距离推理。

(1) 当定性距离满足单调递增约束条件时

——设 $i \geqslant j$，由式(4.8)得

$$(a_i,a_{i+1}]-(a_j,a_{j+1}]=(a_i-a_{j+1},a_{i+1}-a_j)=(\delta_{j+1}+\cdots+\delta_{i-1},a_{i+1}-a_j)$$
$$\subseteq(\delta_0+\cdots+\delta_{i-j-2},a_{i+1}-a_j)\subseteq(a_{i-j-1},a_{i+1}) \tag{4.19}$$

因此，根据定性距离的定义得

$$q_i \wedge q_j \to q_{i-j-1} \vee \cdots \vee q_i \quad (i \geqslant j) \tag{4.20}$$

——设 $i \leqslant j$，由式(4.8)得

$$(a_i,a_{i+1}]-(a_j,a_{j+1}]=(a_i-a_{j+1},a_{i+1}-a_j)$$
$$=(-\delta_i-\delta_{i+1}-\cdots-\delta_j,-\delta_{i+1}-\cdots-\delta_{j-1})$$
$$\subseteq(-a_{j+1},-a_{j-i-1}) \tag{4.21}$$

因此，根据定性距离的定义得

$$q_i \wedge q_j \to q_{j-i-1} \vee \cdots \vee q_j \quad (i \leqslant j) \tag{4.22}$$

根据式(4.20)和式(4.22)，得定性距离推理的组合推理表4.6。

表4.6 单调递增反方向定性推理组合

$\wedge$	q_0	q_1	q_2	q_3	q_4
q_0	q_0	q_0,q_1	q_1,q_2	q_2,q_3	q_3,q_4
q_1	q_0,q_1	q_0,q_1	q_0,q_1,q_2	q_1,q_2,q_3	q_2,q_3,q_4
q_2	q_1,q_2	q_0,q_1,q_2	q_0,q_1,q_2	q_0,q_1,q_2,q_3	q_1,q_2,q_3,q_4
q_3	q_2,q_3	q_1,q_2,q_3	q_0,q_1,q_2,q_3	q_0,q_1,q_2,q_3	q_0,q_1,q_2,q_3,q_4
q_4	q_3,q_4	q_2,q_3,q_4	q_1,q_2,q_3,q_4	q_0,q_1,q_2,q_3,q_4	q_0,q_1,q_2,q_3,q_4

第 4 章 定性距离描述及其推理

(2) 当定性距离满足范围限制约束条件时

——设 $j=i$，则

$$(a_i,a_{i+1}]-(a_i,a_{i+1}]=(a_i-a_{i+1},a_{i+1}-a_i)=(-\delta_i,\delta_i) \quad (4.23)$$

由式(4.3)可知

$$\delta_i \in (a_i,a_i+\delta_i]=(a_i,a_{i+1}] \quad (4.24)$$

由式(4.23)和式(4.24)得

$$(a_i,a_{i+1}]-(a_i,a_{i+1}]\subseteq(-a_{i+1},a_{i+1}) \quad (4.25)$$

根据定性距离的定义得

$$q_i \wedge q_i \to q_0 \vee q_1 \vee \cdots \vee q_i \quad (4.26)$$

——设 $i>j$，则

当 $i=j+1$ 时，根据式(4.8)得

$$(a_i,a_{i+1}]-(a_j,a_{j+1}]=(0,a_{j+2}-a_j)=(0,\delta_j+\delta_{j+1}) \quad (4.27)$$

由范围限制约束条件得

$$a_{j+1} \leqslant \delta_{j+1} \leqslant \delta_j+\delta_{j+1} \leqslant \Delta_{j+1}=a_{j+2} \quad (4.28)$$

根据式(4.27)和式(4.28)得

$$(a_i,a_{i+1}]-(a_j,a_{j+1}]\subseteq(0,a_{j+2})=(0,a_{i+1}) \quad (4.29)$$

根据定性距离的定义可知

$$q_i \wedge q_j \to q_0 \vee q_1 \vee \cdots \vee q_i \quad (4.30)$$

当 $i \geqslant j+2$ 时

$$(a_i,a_{i+1}]-(a_j,a_{j+1}]=(a_i-a_{j+1},a_{i+1}-a_j)=(\delta_{j+1}+\cdots+\delta_{i-1},a_{i+1}-a_j) \quad (4.31)$$

根据式(4.3)得

$$\delta_{j+1}+\cdots+\delta_{i-1} \geqslant \delta_{i-1} \geqslant a_{i-1} \quad (4.32)$$

由式(4.31)和式(4.32)得

$$(a_i,a_{i+1}]-(a_j,a_{j+1}]\subseteq(a_{i-1},a_{i+1}) \quad (4.33)$$

根据定性距离的定义得

$$q_i \wedge q_j \to q_{i-1} \vee q_i \quad (4.34)$$

——设 $j>i$，当 $j=i+1$ 时

$$(a_i,a_{i+1}]-(a_j,a_{j+1}]=(a_i,a_{i+1}]-(a_{i+1},a_{i+2}]=(a_i-a_{i+2},0)=(-\delta_{i+1}-\delta_i,0) \quad (4.35)$$

由式(4.28)和式(4.35)得

$$(a_i,a_{i+1}]-(a_j,a_{j+1}]\subseteq(-a_{i+2},0)=(-a_{j+1},0) \quad (4.36)$$

根据定性距离的定义得

$$q_i \wedge q_j \to q_0 \vee q_1 \vee \cdots \vee q_j \quad (4.37)$$

当 $j \geqslant i+2$ 时

$$(a_i, a_{i+1}] - (a_j, a_{j+1}] = (a_i - a_{j+1}, a_{i+1} - a_j) = (a_i - a_{j+1}, -\delta_{i+1} - \delta_{i+2} - \cdots - \delta_{j-1}) \tag{4.38}$$

$$\delta_{i+1} + \cdots \delta_{j-1} \geqslant \delta_{j-1} \geqslant a_{j-1} \tag{4.39}$$

由式(4.38)和式(4.39)得

$$(a_i, a_{i+1}] - (a_j, a_{j+1}] \subseteq (-a_{j+1}, -a_{j-1}) \tag{4.40}$$

根据定性距离的定义得

$$q_i \wedge q_j \to q_{j-1} \vee q_j \tag{4.41}$$

根据式(4.26)、式(4.34)和式(4.41)得

$$q_i \wedge q_j \to \begin{cases} q_0 \vee \cdots \vee q_{\max\{i,j\}} & |i-j| \leqslant 1 \\ q_{\max\{i,j\}-1} \vee q_{\max\{i,j\}} & |i-j| \geqslant 2 \end{cases} \tag{4.42}$$

根据式(4.42)得定性距离组合推理表 4.7。

表 4.7 当满足范围限制的反方向定性距离推理组合

$\wedge$	q_0	q_1	q_2	q_3	q_4
q_0	q_0	q_0, q_1	q_1, q_2	q_2, q_3	q_3, q_4
q_1	q_0, q_1	q_0, q_1	q_0, q_1, q_2	q_2, q_3	q_3, q_4
q_2	q_1, q_2	q_0, q_1, q_2	q_0, q_1, q_2	q_0, q_1, q_2, q_3	q_3, q_4
q_3	q_2, q_3	q_2, q_3	q_0, q_1, q_2, q_3	q_0, q_1, q_2, q_3	q_0, q_1, q_2, q_3, q_4
q_4	q_3, q_4	q_3, q_4	q_3, q_4	q_0, q_1, q_2, q_3, q_4	q_0, q_1, q_2, q_3, q_4

根据吸收律,当 $\varepsilon = \frac{1}{k^2}$ 时,只要 $\frac{\delta_i}{\delta_j} \leqslant \varepsilon$,就有 $\delta_i + \delta_j \approx q_j$。

$$(a_i, a_{i+1}] - (a_j, a_{j+1}] = (a_i - a_{j+1}, a_{i+1} - a_j) = (-\delta_i - \cdots - \delta_j, -\delta_{i+1} - \cdots - \delta_{j-1})$$
$$\approx (-\delta_0 - \cdots - \delta_j, -\delta_0 - \cdots - \delta_{j-1}) = (-a_{j+1}, -a_j) \tag{4.43}$$

同理

$$(a_j, a_{j+1}] - (a_i, a_{i+1}] \approx (a_j, a_{j+1}) \tag{4.44}$$

根据式(4.43)和式(4.44)可知,当 $|i-j| \geqslant 2$ 时,有

$$q_i \wedge q_j \to q_{\max\{i,j\}} \tag{4.45}$$

根据式(4.42)和式(4.45)得定性距离的组合推理表 4.8。

表 4.8 满足范围限制、$\frac{\delta_i}{\delta_j} \leqslant \frac{1}{k^2}$ 和吸收律的反方向定性距离推理组合

$\wedge$	q_0	q_1	q_2	q_3	q_4
q_0	q_0	q_0, q_1	q_2	q_3	q_4
q_1	q_0, q_1	q_0, q_1	q_0, q_1, q_2	q_3	q_4
q_2	q_2	q_0, q_1, q_2	q_0, q_1, q_2	q_0, q_1, q_2, q_3	q_4
q_3	q_3	q_3	q_0, q_1, q_2, q_3	q_0, q_1, q_2, q_3	q_0, q_1, q_2, q_3, q_4
q_4	q_4	q_4	q_4	q_0, q_1, q_2, q_3, q_4	q_0, q_1, q_2, q_3, q_4

4.2.4 定性推理分析

根据上面的推理过程可以看到,只要定性距离单调递增,表 4.3 至表 4.8 与具体的 k 值无关。在根据吸收律研究定性距离推理时,把二元关系"$\gg$"解释为:对于任意小的正数 $0<\varepsilon<1$,只要 $\delta_i/\delta_j \leqslant \varepsilon$,就有 $\delta_j \gg \delta_i$。表 4.5 和表 4.8 是 $\varepsilon=\dfrac{1}{k^2}$ 的推理结果。如果将"$\gg$"解释为:存在常数 $p>0$,使得只要 $\delta_j-\delta_i \geqslant p$,就有 $\delta_j \gg \delta_i$,那么定性距离推理结果与 k 值和 $p(p>0)$ 值有关。表 4.9 至表 4.11 是满足约束条件(1)~(3)的同向定性距离在不同的 k 和 $p(p>0)$ 下的组合推理表。表 4.12 至表 4.14 是满足约束条件(1)~(3)的反向定性距离在不同的 k 和 $p(p>0)$ 下的组合推理表。根据表 4.9 至表 4.14 可知,对二元算子"$\gg$"的不同解释影响了推理结果。

表 4.9　$k=2, p=20\delta_0$ 时,同方向定性距离推理组合

$\wedge$	q_0	q_1	q_2	q_3	q_4
q_0	q_0, q_1	q_1, q_2	q_2, q_3	q_3, q_4	q_4
q_1	q_1, q_2	q_1, q_2, q_3	q_2, q_3	q_3, q_4	q_4
q_2	q_2, q_3	q_2, q_3	q_2, q_3, q_4	q_3, q_4	q_4
q_3	q_3, q_4	q_3, q_4	q_3, q_4	q_3, q_4	q_4
q_4	q_4	q_4	q_4	q_4	q_4

表 4.10　$k=5, p=20\delta_0$ 时,同方向定性距离推理组合

$\wedge$	q_0	q_1	q_2	q_3	q_4
q_0	q_0, q_1	q_1, q_2	q_2	q_3	q_4
q_1	q_1, q_2	q_1, q_2	q_2	q_3	q_4
q_2	q_2	q_2	q_2, q_3	q_3	q_4
q_3	q_3	q_3	q_3	q_3, q_4	q_4
q_4	q_4	q_4	q_4	q_4	q_4

表 4.11　$k=5, p=30\delta_0$ 时,同方向定性距离推理组合

$\wedge$	q_0	q_1	q_2	q_3	q_4
q_0	q_0, q_1	q_1, q_2	q_2, q_3	q_3	q_4
q_1	q_1, q_2	q_1, q_2	q_2, q_3	q_3	q_4
q_2	q_2, q_3	q_2, q_3	q_2, q_3, q_4	q_3	q_4
q_3	q_3	q_3	q_3	q_3, q_4	q_4
q_4	q_4	q_4	q_4	q_4	q_4

表 4.12 $k=2, p=20\delta_0$ 时,反方向定性距离推理组合

∧	q_0	q_1	q_2	q_3	q_4
q_0	q_0	q_0,q_1	q_1,q_2	q_2,q_3	q_3,q_4
q_1	q_0,q_1	q_0,q_1	q_0,q_1,q_2	q_2,q_3	q_3,q_4
q_2	q_1,q_2	q_0,q_1,q_2	q_0,q_1,q_2	q_0,q_1,q_2,q_3	q_3,q_4
q_3	q_2,q_3	q_2,q_3	q_0,q_1,q_2,q_3	q_0,q_1,q_2,q_3	q_0,q_1,q_2,q_3,q_4
q_4	q_3,q_4	q_3,q_4	q_3,q_4	q_0,q_1,q_2,q_3,q_4	q_0,q_1,q_2,q_3,q_4

表 4.13 $k=5, p=20\delta_0$ 时,反方向定性距离推理组合

∧	q_0	q_1	q_2	q_3	q_4
q_0	q_0	q_0,q_1	q_2	q_3	q_4
q_1	q_0,q_1	q_0,q_1	q_2	q_3	q_4
q_2	q_2	q_2	q_0,q_1,q_2	q_3	q_4
q_3	q_3	q_3	q_3	q_0,q_1,q_2,q_3	q_4
q_4	q_4	q_4	q_4	q_4	q_4

表 4.14 $k=5, p=30\delta_0$ 时,反方向定性距离推理组合

∧	q_0	q_1	q_2	q_3	q_4
q_0	q_0	q_0,q_1	q_1,q_2	q_3	q_4
q_1	q_0,q_1	q_0,q_1	q_0,q_1,q_2	q_3	q_4
q_2	q_1,q_2	q_0,q_1,q_2	q_0,q_1,q_2	q_3	q_4
q_3	q_3	q_3	q_3	q_0,q_1,q_2,q_3	q_4
q_4	q_4	q_4	q_4	q_4	q_4

上述研究的定性推理是参照物 A 和 B 与目标对象 C 在同一条直线上的情况,当三者不在一条直线上时,需要考虑方向关系。二维空间和三维空间中的其他情况将在第 7 章中详细研究。

第 5 章 混合空间关系定性推理

混合空间关系推理是由一种空间关系推导另一种空间关系。如由 A 与 B 的拓扑关系 $top(A,B)$ 以及 B 与 C 的拓扑关系 $top(B,C)$,推导方向关系 $dir(A,B)$,简称由拓扑关系推导方向关系。拓扑关系(top)、方向关系(dir)、定性距离($qdis$)间的混合推理共有以下 6 种类型(Sharma,1996)

$$top(A,B) \wedge top(B,C) \rightarrow dir(A,C)$$
$$top(A,B) \wedge top(B,C) \rightarrow qdis(A,C)$$
$$dir(A,B) \wedge dir(B,C) \rightarrow top(A,C)$$
$$dir(A,B) \wedge dir(B,C) \rightarrow qdis(A,C)$$
$$qdis(A,B) \wedge qdis(B,C) \rightarrow dir(A,C)$$
$$qdis(A,B) \wedge qdis(B,C) \rightarrow top(A,C)$$

本章用区间关系对描述方向区域和经常使用的 8 种拓扑关系,研究拓扑关系与方向关系的混合推理。

§5.1 二维混合空间关系定性推理

在第 3 章中,采用投影法描述目标对象与参照物间的方向关系。投影方法是指把空间对象分别向 X 轴和 Y 轴投影,利用投影后的坐标来描述方向关系。参照物 $A(x_A,y_A)$ 在 X 轴上的投影,将 X 轴分为 3 个部分:

$\{(x,y)|x<\min\{x_A|(x_A,y_A)\in A\}\}$

$\{(x,y)|\min\{x_A|(x_A,y_A)\in A\}\leqslant x\leqslant\max\{x_A|(x_A,y_A)\in A\}\}$

$\{(x,y)|x>\max\{x_A|(x_A,y_A)\in A\}\}$

参照物在 Y 轴的投影,将 Y 轴分为 3 个部分:

$\{(x,y)|y<\min\{y_A|(x_A,y_A)\in A\}\}$

$\{(x,y)|\min\{y_A|(x_A,y_A)\in A\}\leqslant y\leqslant\max\{y_A|(x_A,y_A)\in A\}\}$

$\{(x,y)|y>\max\{y_A|(x_A,y_A)\in A\}\}$

图 5.1 表示 Allen 的一维区间关系(Allen,1983;Sharma,1996)。设目标对

象为 $B(x_B,y_B)$，当 $(x_B,y_B)\in\{(x,y)|x<\min\{x_A|(x_A,y_A)\in A\}\}$ 时，目标对象 B 与参照物 A 在 X 轴上的 Allen 的区间关系描述为 after 或 metby；当 $(x_B,y_B)\in \{(x,y)|\min\{x_A|(x_A,y_A)\in A\}\leqslant x\leqslant\max\{x_A|(x_A,y_A)\in A\}\}$ 时，目标对象 B 与参照物 A 在 X 轴上的 Allen 的区间关系描述为 startedby, contain, equal 或 finishedby；当 $(x_B,y_B)\in\{(x,y)|x>\max\{x_A|(x_A,y_A)\in A\}\}$ 时，目标对象 B 与参照物 A 在 X 轴上的 Allen 区间关系描述为 before 或 meet。同样，当 B 的坐标范围分别为 $\{(x,y)|y<\min\{y_A|(x_A,y_A)\in A\}\}$、$\{(x,y)|\min\{y_A|(x_A,y_A)\in A\}\leqslant y\leqslant\max\{y_A|(x_A,y_A)\in A\}\}$、$\{(x,y)|y>\max\{y_A|(x_A,y_A)\in A\}\}$ 时，目标对象 B 与参照物 A 在 Y 轴上的 Allen 区间关系分别描述为 after 或 metby, startedby, contain, equal, finishedby 和 before 或 meet。

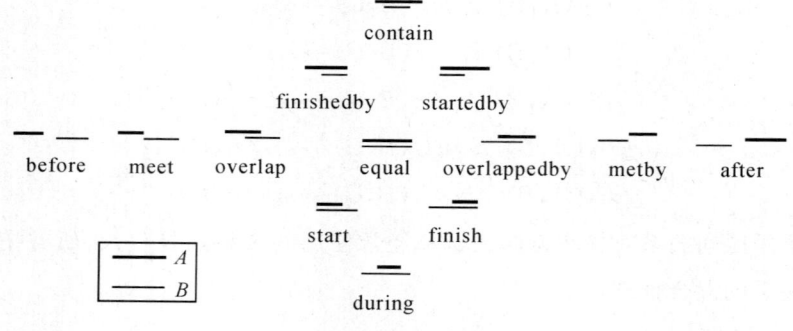

图 5.1　Allen 的一维区间关系

当 $(x_B,y_B)\in\{(x,y)|x<\min\{x_A|(x_A,y_A)\in A\},y>\max\{y_A|(x_A,y_A)\in A\}\}$ 时，用 Allen 的区间关系对描述为 (after, before)(metby, before),(metby, meet),(after, meet)，如图 5.2(a) 所示。当 $(x_B,y_B)\in\{(x,y)|\min\{x_A|(x_A,y_A)\in A\}\leqslant x\leqslant\max\{x_A|(x_A,y_A)\in A\},y>\max\{y_A|(x_A,y_A)\in A\}\}$ 时，用 Allen 的区间关系对描述为 (startedby, meet),(contain, meet),(finishedby, meet),(equal, meet),(startedby, before),(contain, before),(finishedby, before),(equal, before)，如图 5.2(b) 所示。

在二维空间中，用区间关系对 (R_1,R_2) 描述目标对象 B 与参照物 A 间的方向关系，其中，R_1 表示目标对象 B 与参照物 A 在 X 轴上的 Allen 区间关系，R_2 表示目标对象 B 与参照物 A 在 Y 轴上的 Allen 区间关系，式(5.1)至式(5.9)是用 Allen 区间关系对描述二维空间的方向区域。其中，$IA=\{\text{before, meet, overlap, finishedby, contain, startedby, start, finish, during, equal, overlappedby, metby, after}\}$，$I=\{\text{finishedby, contain, startedby, equal}\}$。

第 5 章 混合空间关系定性推理

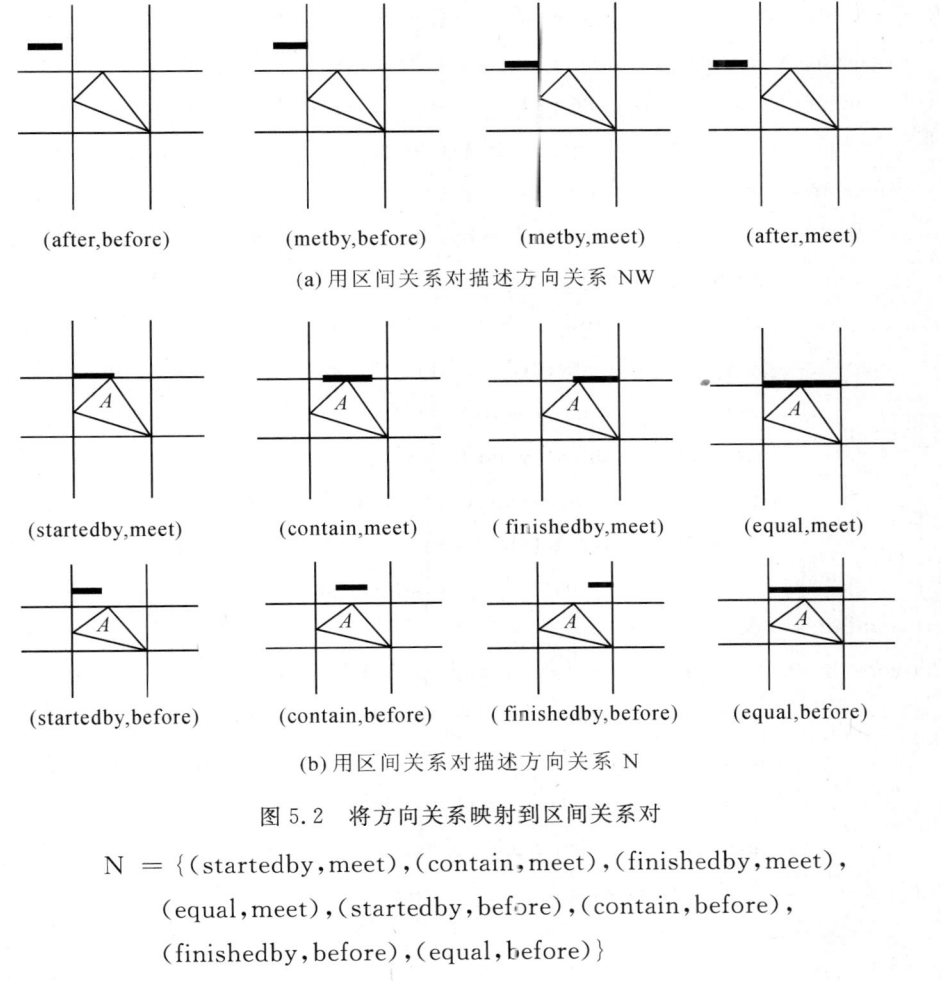

图 5.2 将方向关系映射到区间关系对

$$N = \{(\text{startedby}, \text{meet}), (\text{contain}, \text{meet}), (\text{finishedby}, \text{meet}),$$
$$(\text{equal}, \text{meet}), (\text{startedby}, \text{before}), (\text{contain}, \text{before}),$$
$$(\text{finishedby}, \text{before}), (\text{equal}, \text{before})\}$$
$$= I \times \{\text{before}, \text{meet}\} \qquad (5.1)$$
$$NE = \{\text{before}, \text{meet}\} \times \{\text{before}, \text{meet}\} \qquad (5.2)$$
$$E = \{\text{before}, \text{meet}\} \times I \qquad (5.3)$$
$$SE = \{\text{before}, \text{meet}\} \times \{\text{metby}, \text{after}\} \qquad (5.4)$$
$$S = I \times \{\text{after}, \text{metby}\} \qquad (5.5)$$
$$SW = \{\text{after}, \text{metby}\} \times \{\text{after}, \text{metby}\} \qquad (5.6)$$
$$W = \{\text{after}, \text{metby}\} \times I \qquad (5.7)$$
$$NW = \{\text{after}, \text{metby}\} \times \{\text{before}, \text{meet}\} \qquad (5.8)$$
$$\text{same} = I \times I \qquad (5.9)$$

根据 Allen 区间关系的定义,定义在二维空间中的区间关系对 (R_1, R_2), R,

$R_2 \in IA$ 与经常使用的 8 种拓扑关系的关系为

$$\text{disjoint} = \{\text{before}, \text{after}\} \times IA \cup IA \times \{\text{before}, \text{after}\} \tag{5.10}$$

$$\text{meet} = \{\text{meet}, \text{metby}\} \times (IA - \{\text{before}, \text{after}\}) \cup$$
$$(IA - \{\text{before}, \text{after}\}) \times \{\text{meet}, \text{metby}\} \tag{5.11}$$

$$\text{overlap} = \{\text{overlap}, \text{overlappedby}\} \times (IA - \{\text{before}, \text{meet}, \text{metby}, \text{after}\}) \cup$$
$$(IA - \{\text{before}, \text{meet}, \text{metby}, \text{after}\}) \times \{\text{overlap}, \text{overlappedby}\} \cup$$
$$\{\text{startedby}, \text{contain}, \text{finishedby}\} \times \{\text{start}, \text{during}, \text{finish}\} \cup$$
$$\{\text{start}, \text{during}, \text{finish}\} \times \{\text{startedby}, \text{contain}, \text{finishedby}\} \tag{5.12}$$

$$\text{cover} = \{\text{startedby}, \text{finishedby}\} \times I \cup I \times \{\text{startedby}, \text{finishedby}\} \cup$$
$$\{\text{contain}\} \times \{\text{startedby}, \text{finishedby}, \text{equal}\} \cup$$
$$\{\text{startedby}, \text{finishedby}, \text{equal}\} \times \{\text{contain}\} \cup$$
$$\{\text{equal}\} \times \{\text{startedby}, \text{contain}, \text{finishedby}\} \cup$$
$$\{\text{startedby}, \text{contain}, \text{finishedby}\} \times \{\text{equal}\}$$
$$= I \times I - \{(\text{contain}, \text{contain}), (\text{equal}, \text{equal})\} \tag{5.13}$$

$$\text{contain} = (\text{contain}, \text{contain}) \tag{5.14}$$

$$\text{coveredby} = \{\text{start}, \text{finish}\} \times \{\text{start}, \text{during}, \text{finish}, \text{equal}\} \cup$$
$$\{\text{start}, \text{during}, \text{finish}, \text{equal}\} \times \{\text{start}, \text{finish}\} \cup$$
$$\{\text{equal}\} \times \{\text{start}, \text{during}, \text{finish}\} \cup \{\text{start}, \text{during}, \text{finish}\} \times \{\text{equal}\} \cup$$
$$\{\text{during}\} \times \{\text{start}, \text{finish}, \text{equal}\} \cup \{\text{start}, \text{finish}, \text{equal}\} \times \{\text{during}\}$$
$$= \{\text{start}, \text{during}, \text{finish}, \text{equal}\} \times \{\text{start}, \text{during}, \text{finish}, \text{equal}\} -$$
$$\{(\text{equal}, \text{equal}), (\text{during}, \text{during})\} \tag{5.15}$$

$$\text{inside} = (\text{during}, \text{during}) \tag{5.16}$$

$$\text{equal} = (\text{equal}, \text{equal}) \tag{5.17}$$

由式(5.1)至式(5.17)可知,根据目标对象 B 与参照物 A 的方向关系,目标对象 C 与参照物 B 的方向关系,推导 C 与 A 的拓扑关系,用式

$$dir(A, B) \wedge dir(B, C) \rightarrow top(A, C) \tag{5.18}$$

表示根据方向关系 $dir(A, B)$ 和 $dir(B, C)$ 推导拓扑关系 $top(A, C)$。

在二维空间中,设 $dir(A, B) = R_1 \times R_2, dir(B, C) = S_1 \times S_2, R_1, R_2, S_1, S_2 \subseteq IA$,那么

$$dir(A, B) \wedge dir(B, C) = R_1 \times R_2 \wedge S_1 \times S_2$$
$$= \{r_{11}, r_{12}, \cdots, r_{1m}\} \times \{r_{21}, r_{22}, \cdots, r_{2n}\} \wedge \{s_{11}, s_{12}, \cdots, s_{1l}\} \times \{s_{21}, s_{22}, \cdots, s_{2s}\}$$
$$= \{(r_{11}, r_{21}), (r_{11}, r_{22}), \cdots, (r_{11}, r_{2n}), (r_{12}, r_{21}), (r_{12}, r_{22}), \cdots, (r_{12}, r_{2n}), \cdots,$$
$$(r_{1m}, r_{21}), (r_{1m}, r_{22}), \cdots, (r_{1m}, r_{2n})\} \wedge \{(s_{11}, s_{21}), (s_{11}, s_{22}), \cdots,$$

$(s_{11},s_{2s}),(s_{12},s_{21}),(s_{12},s_{22}),\cdots,(s_{12},s_{2s}),(s_{1l},s_{21}),(s_{1l},s_{22}),\cdots,(s_{1l},s_{2s})\}$

$= (r_{11},r_{21}) \wedge (s_{11},s_{21}) \bigcup (r_{11},r_{21}) \wedge (s_{11},s_{22}) \bigcup \cdots \bigcup (r_{11},r_{21}) \wedge (s_{1l},s_{2s}) \bigcup \cdots \bigcup$

$(r_{1m},r_{2n}) \wedge (s_{11},s_{21}) \bigcup (r_{1m},r_{2n}) \wedge (s_{11},s_{22}) \bigcup \cdots \bigcup (r_{1m},r_{2n}) \wedge (s_{1l},s_{2s})$

$= (r_{11} \wedge s_{11}, r_{21} \wedge s_{21}) \bigcup (r_{11} \wedge s_{11}, r_{21} \wedge s_{22}) \bigcup \cdots \bigcup (r_{11} \wedge s_{11}, r_{21} \wedge s_{2s}) \bigcup \cdots \bigcup$

$(r_{1m} \wedge s_{11}, r_{2n} \wedge s_{21}) \bigcup (r_{1m} \wedge s_{11}, r_{2n} \wedge s_{22}) \bigcup \cdots \bigcup (r_{1m} \wedge s_{11}, r_{2n} \wedge s_{2s})$

$= \{r_{i_1} \wedge s_{j_1} | r_{i_1} \in R_1, s_{j_1} \in S_1\} \times \{r_{i_2} \wedge s_{j_2} | r_{i_2} \in R_2, s_{j_2} \in S_2\}$

$= (R_1 \wedge S_1) \times (R_2 \wedge S_2)$ (5.19)

式中,$1 \leqslant m,n,l,s \leqslant 13$。

Allen 的 13 个一维区间关系用图 5.3 表示(Sharma,1996)。根据 Allen 区间关系的定义,由 A 与 B 的区间关系以及 B 与 C 的区间关系,可以推导 A 与 C 的区间关系,推理结果见表 5.1。

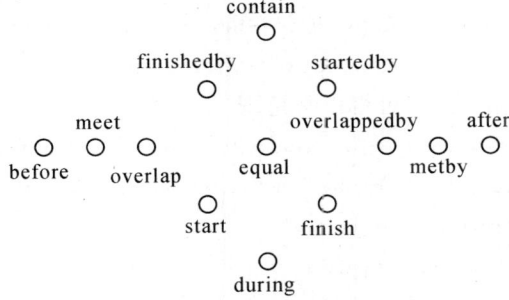

图 5.3 描述 Allen 区间关系的图标

表 5.1 Allen 区间关系推理组合

在根据区间关系进行定性推理时,下列区间关系推理满足传递性。

$$\text{before} \wedge \text{before} \rightarrow \text{before} \quad (5.20)$$

$$\text{contain} \wedge \text{contain} \rightarrow \text{contain} \quad (5.21)$$
$$\text{finishedby} \wedge \text{finishedby} \rightarrow \text{finishedby} \quad (5.22)$$
$$\text{startedby} \wedge \text{startedby} \rightarrow \text{startedby} \quad (5.23)$$
$$\text{equal} \wedge \text{equal} \rightarrow \text{equal} \quad (5.24)$$
$$\text{start} \wedge \text{start} \rightarrow \text{start} \quad (5.25)$$
$$\text{finish} \wedge \text{finish} \rightarrow \text{finish} \quad (5.26)$$
$$\text{during} \wedge \text{during} \rightarrow \text{during} \quad (5.27)$$
$$\text{after} \wedge \text{after} \rightarrow \text{after} \quad (5.28)$$

仅有下列4个区间关系推理不满足传递性,即

$$\text{meet} \wedge \text{meet} \rightarrow \text{before} \quad (5.29)$$
$$\text{metby} \wedge \text{metby} \rightarrow \text{after} \quad (5.30)$$
$$\text{overlap} \wedge \text{overlap} \rightarrow \text{before} \vee \text{meet} \vee \text{overlap} \quad (5.31)$$
$$\text{overlappedby} \wedge \text{overlappedby} \rightarrow \text{after} \vee \text{metby} \vee \text{overlappedby} \quad (5.32)$$

根据 Allen 区间关系的定义,可得以下定理

$$I \wedge \{\text{before},\text{meet}\} \rightarrow \{\text{before},\text{meet},\text{overlap},\text{finishedby},\text{contain}\} \quad (5.33)$$
$$I \wedge \{\text{metby},\text{after}\} \rightarrow \{\text{after},\text{metby},\text{overlappedby},\text{startedby},\text{contain}\} \quad (5.34)$$
$$\{\text{before},\text{meet}\} \wedge \{\text{before},\text{meet}\} \rightarrow \{\text{before}\} \quad (5.35)$$
$$\{\text{metby},\text{after}\} \wedge \{\text{metby},\text{after}\} \rightarrow \{\text{after}\} \quad (5.36)$$
$$\{\text{before},\text{meet}\} \wedge I \rightarrow \{\text{before},\text{meet}\} \quad (5.37)$$
$$\{\text{metby},\text{after}\} \wedge I \rightarrow \{\text{after},\text{metby}\} \quad (5.38)$$
$$I \wedge I \rightarrow I \quad (5.39)$$

根据式(5.1)至式(5.19),可由方向关系推导拓扑关系,举例如下。

例 5.1

$$O_{SW} \wedge O_N = \{\text{after},\text{metby}\} \times \{\text{after},\text{metby}\} \wedge I \times \{\text{before},\text{meet}\}$$
$$= \{\text{after},\text{metby}\} \wedge I \times \{\text{after},\text{metby}\} \wedge \{\text{before},\text{meet}\}$$
$$\rightarrow \{\text{after},\text{metby}\} \times IA$$
$$\rightarrow \text{disjoint} \vee \text{meet}$$

$$O_{NW} \wedge O_E = \{\text{after},\text{metby}\} \times \{\text{before},\text{meet}\} \wedge I \times \{\text{before},\text{meet}\}$$
$$= \{\text{after},\text{metby}\} \wedge I \times \{\text{after},\text{metby}\} \times \{\text{before},\text{meet}\}$$
$$\rightarrow IA \times \{\text{before},\text{meet}\}$$
$$\rightarrow \text{disjoint} \vee \text{meet}$$

$$O_{same} \wedge O_N = I \times I \wedge I \times \{\text{before},\text{meet}\}$$
$$= I \wedge I \times I \times \{\text{before},\text{meet}\}$$
$$\rightarrow I \times \{\text{contain},\text{finishedby},\text{overlap},\text{meet},\text{before}\}$$
$$\rightarrow \text{disjoint} \vee \text{meet} \vee \text{overlap} \vee \text{cover} \vee \text{contain}$$

第 5 章 混合空间关系定性推理

$$O_{\text{same}} \wedge O_{\text{NW}} = I \times I \wedge \{\text{after}, \text{metby}\} \times \{\text{before}, \text{meet}\}$$
$$= I \wedge \{\text{after}, \text{metby}\} \times I \wedge \{\text{before}, \text{meet}\}$$
$$\rightarrow \{\text{after}, \text{metby}, \text{overlappedby}, \text{startedby}, \text{contain}\} \times$$
$$\{\text{contain}, \text{finishedby}, \text{overlap}, \text{meet}, \text{before}\}$$
$$\rightarrow \text{disjoint} \vee \text{meet} \vee \text{overlap} \vee \text{cover} \vee \text{contain}$$

$$O_{\text{same}} \wedge O_{\text{same}} = I \times I \wedge I \times I \rightarrow I \times I$$
$$\rightarrow \text{cover} \vee \text{contain} \vee \text{equal}$$

图 5.4 表示 8 种拓扑关系(Sharma,1996；Egenhofer et al,1992)。根据方向区域的定义以及拓扑关系的定义可得定性推理 $dir(A,B) \wedge dir(B,C) \rightarrow top(A,C)$，具体结果见表 5.2。

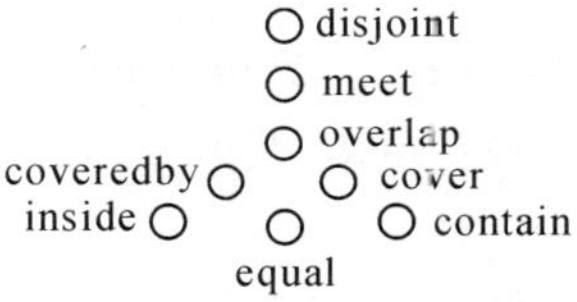

图 5.4 拓扑关系图标

表 5.2 二维空间基于方向关系推导拓扑关系

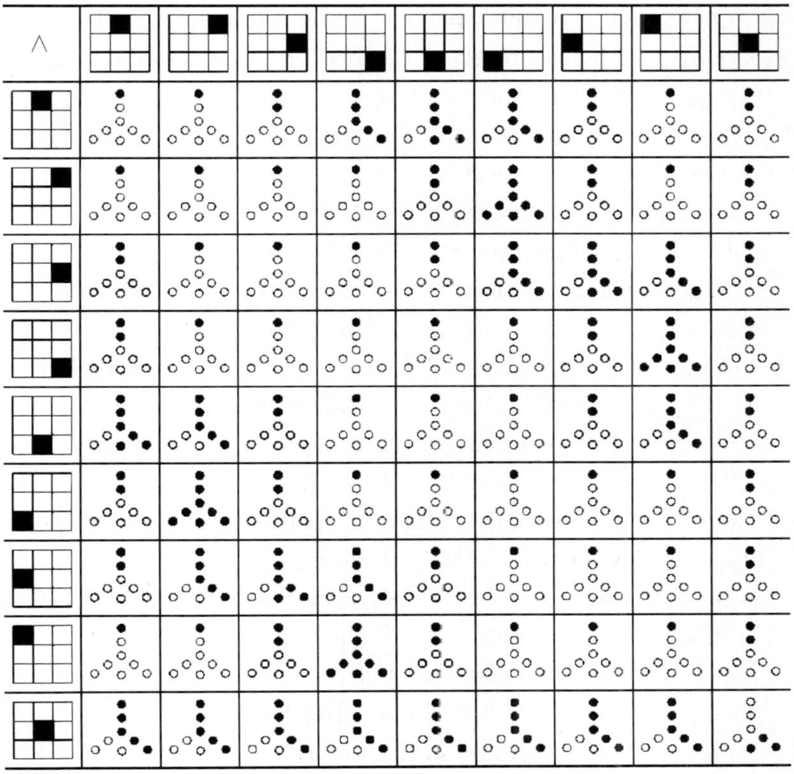

§5.2 三维混合空间关系定性推理

5.2.1 拓扑关系与方向关系

在三维空间中,参照物 A 在 Z 坐标轴上的投影将 Z 轴划分为 3 个部分:
$\text{up} = \{(x,y,z) \mid z > \max\{z_A \mid (x_A,y_A,z_A) \in A\}\}$
$\text{between} = \{(x,y,z) \mid \min\{z_A \mid (x_A,y_A,z_A) \in A\} \leqslant z \leqslant \max\{z_A \mid (x_A,y_A,z_A) \in A\}\}$
$\text{down} = \{(x,y,z) \mid z < \min\{z_A \mid (x_A,y_A,z_A) \in A\}\}$

设 B 点坐标为 (x_B,y_B,z_B),当 $(x_B,y_B,z_B) \in \text{down}$ 时,B 与 A 在 Z 轴上的区间关系为 after 或 metby;当 $(x_B,y_B,z_B) \in \text{between}$ 时,B 与 A 在 Z 轴上的区间关系为 startedby,contain,equal 或 finishedby;当 $(x_B,y_B,z_B) \in \text{up}$ 时,B 与 A 在 Z 轴上的区间关系为 before 或 meet。

在三维空间中,用 (R_1,R_2,R_3)(其中 $R_1,R_2,R_3 \in IA$)描述方向区域如下。

$$O_{\text{N,up}} = I \times \{\text{before,meet}\} \times \{\text{before,meet}\} \tag{5.40}$$

$$O_{\text{N,between}} = I \times \{\text{before,meet}\} \times I \tag{5.41}$$

$$O_{\text{N,down}} = I \times \{\text{before,meet}\} \times \{\text{after,metby}\} \tag{5.42}$$

$$O_{\text{NE,up}} = \{\text{before,meet}\} \times \{\text{before,meet}\} \times \{\text{before,meet}\} \tag{5.43}$$

$$O_{\text{NE,between}} = \{\text{before,meet}\} \times \{\text{before,meet}\} \times I \tag{5.44}$$

$$O_{\text{NE,down}} = \{\text{before,meet}\} \times \{\text{before,meet}\} \times \{\text{after,metby}\} \tag{5.45}$$

$$O_{\text{E,up}} = \{\text{before,meet}\} \times I \times \{\text{before,meet}\} \tag{5.46}$$

$$O_{\text{E,between}} = \{\text{before,meet}\} \times I \times I \tag{5.47}$$

$$O_{\text{E,down}} = \{\text{before,meet}\} \times I \times \{\text{after,metby}\} \tag{5.48}$$

$$O_{\text{SE,up}} = \{\text{before,meet}\} \times \{\text{metby,after}\} \times \{\text{before,meet}\} \tag{5.49}$$

$$O_{\text{SE,between}} = \{\text{before,meet}\} \times \{\text{metby,after}\} \times I \tag{5.50}$$

$$O_{\text{SE,down}} = \{\text{before,meet}\} \times \{\text{metby,after}\} \times \{\text{after,metby}\} \tag{5.51}$$

$$O_{\text{S,up}} = I \times \{\text{after,metby}\} \times \{\text{before,meet}\} \tag{5.52}$$

$$O_{\text{S,between}} = I \times \{\text{after,metby}\} \times I \tag{5.53}$$

$$O_{\text{S,down}} = I \times \{\text{after,metby}\} \times \{\text{after,metby}\} \tag{5.54}$$

$$O_{\text{SW,up}} = \{\text{after,metby}\} \times \{\text{after,metby}\} \times \{\text{before,meet}\} \tag{5.55}$$

$$O_{\text{SW,between}} = \{\text{after,metby}\} \times \{\text{after,metby}\} \times I \tag{5.56}$$

$$O_{\text{SW,down}} = \{\text{after,metby}\} \times \{\text{after,metby}\} \times \{\text{after,metby}\} \tag{5.57}$$

$$O_{\text{W,up}} = \{\text{after,metby}\} \times I \times \{\text{before,meet}\} \tag{5.58}$$

$$O_{\text{W,between}} = \{\text{after,metby}\} \times I \times I \tag{5.59}$$

$$O_{\text{W,down}} = \{\text{after,metby}\} \times I \times \{\text{after,metby}\} \tag{5.60}$$

$$O_{\text{NW,up}} = \{\text{after,metby}\} \times \{\text{before,meet}\} \times \{\text{before,meet}\} \tag{5.61}$$

$$O_{\text{NW,between}} = \{\text{after},\text{metby}\} \times \{\text{before},\text{meet}\} \times I \tag{5.62}$$

$$O_{\text{NW,down}} = \{\text{after},\text{metby}\} \times \{\text{before},\text{meet}\} \times \{\text{after},\text{metby}\} \tag{5.63}$$

$$O_{\text{same,up}} = I \times I \times \{\text{before},\text{meet}\} \tag{5.64}$$

$$O_{\text{same,between}} = I \times I \times I \tag{5.65}$$

$$O_{\text{same,down}} = I \times I \times \{\text{after},\text{metby}\} \tag{5.66}$$

根据 Allen 区间关系的定义,在三维空间中区间关系对 (R_1, R_2, R_3)(其中 R_1, $R_2, R_3 \in IA$)与经常使用的 8 种拓扑关系的关系为

$$\begin{aligned}\text{disjoint} = &\{\text{before},\text{after}\} \times IA \times IA \cup IA \times \{\text{before},\text{after}\} \times IA \cup IA \times \\ & IA \times \{\text{before},\text{after}\}\end{aligned} \tag{5.67}$$

$$\begin{aligned}\text{meet} = &\{\text{meet},\text{metby}\} \times (IA - \{\text{before},\text{after}\}) \times (IA - \{\text{before},\text{after}\}) \cup \\ & (IA - \{\text{before},\text{after}\}) \times \{\text{meet},\text{metby}\} \times (IA - \{\text{before},\text{after}\}) \cup \\ & (IA - \{\text{before},\text{after}\}) \times (IA - \{\text{before},\text{after}\}) \times \{\text{meet},\text{metby}\}\end{aligned} \tag{5.68}$$

令 $OV = IA - \{\text{before},\text{meet},\text{metby},\text{after}\}$,那么

$$\begin{aligned}\text{overlap} = & \{\text{overlap},\text{overlappedby}\} \times OV \times OV \cup OV \times \{\text{overlap},\text{overlappedby}\} \times \\ & OV \cup OV \times OV \times \{\text{overlap},\text{overlappedby}\} \cup \{\text{startedby},\text{contain}, \\ & \text{finishedby}\} \times \{\text{start},\text{during},\text{finish}\} \times OV \cup \{\text{startedby},\text{contain}, \\ & \text{finishedby}\} \times OV \times \{\text{start},\text{during},\text{finish}\} \cup \{\text{start},\text{during},\text{finish}\} \times OV \times \\ & \{\text{start},\text{during},\text{finish}\} \cup \{\text{start},\text{during},\text{finish}\} \times \{\text{start},\text{during},\text{finish}\} \times \\ & OV \cup OV \times \{\text{startedby},\text{contain},\text{finishedby}\} \times \{\text{start},\text{during},\text{finish}\} \cup \\ & OV \times \{\text{start},\text{during},\text{finish}\} \times \{\text{startedby},\text{contain},\text{finishedby}\}\end{aligned} \tag{5.69}$$

$$\text{cover} = I \times I \times I - \{(\text{contain},\text{contain},\text{contain}), (\text{equal},\text{equal},\text{equal})\} \tag{5.70}$$

$$\text{contain} = (\text{contain},\text{contain},\text{contain}) \tag{5.71}$$

$$\begin{aligned}\text{coveredby} = & \{\text{start},\text{finish}\} \times \{\text{start},\text{during},\text{finish},\text{equal}\} \cup \\ & \{\text{start},\text{during},\text{finish},\text{equal}\} \times \{\text{start},\text{finish}\} \cup \\ & \{\text{equal}\} \times \{\text{start},\text{during},\text{finish}\} \cup \{\text{start},\text{during},\text{finish}\} \times \{\text{equal}\} \\ & \{\text{during}\} \times \{\text{start},\text{finish},\text{equal}\} \cup \{\text{start},\text{finish},\text{equal}\} \times \{\text{during}\} \\ = & \{\text{start},\text{during},\text{finish},\text{equal}\} \times \{\text{start},\text{during},\text{finish},\text{equal}\} \times \\ & \{\text{start},\text{during},\text{finish},\text{equal}\} - \{(\text{equal},\text{equal},\text{equal}), \\ & (\text{during},\text{during},\text{during})\}\end{aligned} \tag{5.72}$$

$$\text{inside} = (\text{during},\text{during},\text{during}) \tag{5.73}$$

$$\text{equal} = (\text{equal},\text{equal},\text{equal}) \tag{5.74}$$

5.2.2 由方向关系推导拓扑关系

在三维空间中,设 $dir(A,B) = R_1 \times R_2 \times R_3$, $dir(B,C) = S_1 \times S_2 \times S_3$, R_1, R_2, $R_3, S_1, S_2, S_3 \subseteq IA$,那么

$$\begin{aligned}dir(A,B) \wedge dir(B,C) &= R_1 \times R_2 \times R_3 \wedge S_1 \times S_2 \times S_3 \\ &= (R_1 \wedge S_1) \times (R_2 \wedge S_2) \times (R_3 \wedge S_3)\end{aligned} \tag{5.75}$$

根据拓扑关系的定义,可得如下定理

$IA \times IA \times I \to$ disjoint $\vee$ meet $\vee$ overlap $\vee$ cover $\vee$ contain $\vee$ equal $\vee$
  coveredby $\vee$ inside   (5.76)

$IA \times I \times I \to$ disjoint $\vee$ meet $\vee$ overlap $\vee$ cover $\vee$ contain $\vee$ equal   (5.77)

$I \times I \times I \to$ cover $\vee$ contain $\vee$ equal   (5.78)

{before, meet, overlap, finisheby, contain} $\times IA \times IA$
  $\to$ disjoint $\vee$ meet $\vee$ overlap $\vee$ cover $\vee$ contain   (5.79)

{before, meet, overlap, finisheby, contain} $\times I \times I$
  $\to$ disjoint $\vee$ meet $\vee$ overlap $\vee$ cover $\vee$ contain   (5.80)

证明：

——式(5.76)的证明如下。

$IA \times IA \times I = \{\{$before, after$\} \cup \{$meet, metby$\} \cup \{$overlap, overlappedby$\} \cup$
  $\{$finishedby, startedby$\} \cup \{$start, finish$\} \cup \{$during$\} \cup \{$equal$\} \cup$
  $\{$contain$\}\} \times IA \times I$

$= \{$before, after$\} \times IA \times I \cup \{$meet, metby$\} \times IA \times I \cup$
  $\{$overlap, overlappedby$\} \times IA \times I \cup \{$finishedby, startedby$\} \times$
  $IA \times I \cup \{$start, finish$\} \times IA \times I \cup \{$during$\} \times IA \times I \cup \{$equal$\} \times$
  $IA \times I \cup \{$contain$\} \times IA \times I$

$=$ disjoint $\vee$ meet $\vee$ overlap $\vee$ cover $\vee$ contain $\vee$ coveredby $\vee$ inside $\vee$ equal

——式(5.77)的证明如下。

$IA \times I \times I = \{\{$before, after$\} \cup \{$meet, metby$\} \cup \{$overlap, overlappedby$\} \cup$
  $\{$finishedby, startedby$\} \cup \{$start, finish$\} \cup \{$during$\} \cup \{$equal$\} \cup$
  $\{$contain$\}\} \times I \times I$

$= \{$before, after$\} \times I \times I \cup \{$meet, metby$\} \times I \times I \cup$
  $\{$overlap, overlappedby$\} \times I \times I \cup \{$finishedby, startedby$\} \times I \times I \cup$
  $\{$start, finish, during$\} \times I \times I \cup \{$equal$\} \times I \times I \cup \{$contain$\} \times I \times I$

$=$ disjoint $\vee$ meet $\vee$ overlap $\vee$ cover $\vee$ contain $\vee$ equal

——式(5.78)的证明如下。

根据 cover(5.70)的定义，可得 $I \times I \times I \to$ cover $\vee$ contain $\vee$ equal。

——式(5.79)的证明如下。

{before, meet, overlap, finisheby, contain} $\times IA \times IA$
  $= \{$before$\} \times IA \times IA \cup \{$meet$\} \times IA \times IA \cup \{$overlap$\} \times IA \times IA \cup$
  $\{$finisheby$\} \times IA \times IA \cup \{$contain$\} \times IA \times IA$

  $\to$ disjoint $\vee$ meet $\vee$ overlap $\vee$ cover $\vee$ contain

——式(5.80)的证明如下。

{before, meet, overlap, finisheby, contain} $\times I \times I$
  $= \{$before$\} \times I \times I \cup \{$meet$\} \times I \times I \cup \{$overlap$\} \times I \times I \cup \{$finisheby$\} \times I \times I \cup$
  $\{$contain$\} \times I \times I$

  $\to$ disjoint $\vee$ meet $\vee$ overlap $\vee$ cover $\vee$ contain

第 5 章　混合空间关系定性推理

例 5.2

$O_{N,up} \wedge O_{NE,up} = I \times \{before, meet\} \times \{before, meet\} \wedge \{before, meet\} \times \{before, meet\} \times \{before, meet\}$

$= I \wedge \{before, meet\} \times \{before, meet\} \wedge \{before, meet\} \times \{before, meet\} \wedge \{before, meet\}$

$\rightarrow \{before, meet, overlap, finisheby, contain\} \times \{before\} \times \{before\}$

$\rightarrow disjoint$

$O_{SE,between} \wedge O_{NE,up} = \{before, meet\} \times \{after, metby\} \times I \wedge \{before, meet\} \times \{before, meet\} \times \{before, meet\}$

$= \{before, meet\} \wedge \{before, meet\} \times \{after, metby\} \wedge \{before, meet\} \times I \wedge \{before, meet\}$

$\rightarrow \{before\} \times IA \times \{before, meet, overlap, finishedby, contain\}$

$\rightarrow disjoint$

$O_{N,up} \wedge O_{E,down} = I \times \{before, meet\} \times \{before, meet\} \wedge \{before, meet\} \times I \times \{after, metby\}$

$= I \wedge \{before, meet\} \times \{before, meet\} \wedge I \times \{before, meet\} \wedge \{after, metby\}$

$\rightarrow \{before, meet, overlap, finisheby, contain\} \times \{before, meet\} \times IA$

$\rightarrow disjoint \vee meet$

$O_{N,up} \wedge O_{SE,down} = I \times \{before, meet\} \times \{before, meet\} \wedge \{before, meet\} \times \{after, metby\} \times \{after, metby\}$

$= I \wedge \{before, meet\} \times \{before, meet\} \wedge \{after, metby\} \times \{before, meet\} \wedge \{after, metby\}$

$\rightarrow \{before, meet, overlap, finisheby, contain\} \times IA \times IA$

$\rightarrow disjoint \vee meet \vee overlap \vee cover \vee contain$

$O_{NE,up} \wedge O_{SW,down} = \{before, meet\} \times \{before, meet\} \times \{before, meet\} \wedge \{after, metby\} \times \{after, metby\} \times \{after, metby\}$

$= \{before, meet\} \wedge \{after, metby\} \times \{before, meet\} \wedge \{after, metby\} \times \{before, meet\} \wedge \{after, metby\}$

$\rightarrow IA \times IA \times IA$

$\rightarrow disjoint \vee meet \vee overlap \vee cover \vee contain \vee coveredby \vee inside \vee equal$

$O_{NE,between} \wedge O_{SW,between} = \{before, meet\} \times \{before, meet\} \times I \wedge \{after, metby\} \times \{after, metby\} \times I$

$= IA \times IA \times I$

$\rightarrow disjoint \vee meet \vee overlap \vee cover \vee contain \vee coveredby \vee inside \vee equal$

根据方向关系推导拓扑关系的所有推理结果见表 5.3。

表 5.3 三维空间基于方向

第 5 章 混合空间关系定性推理 107

关系推导拓扑关系

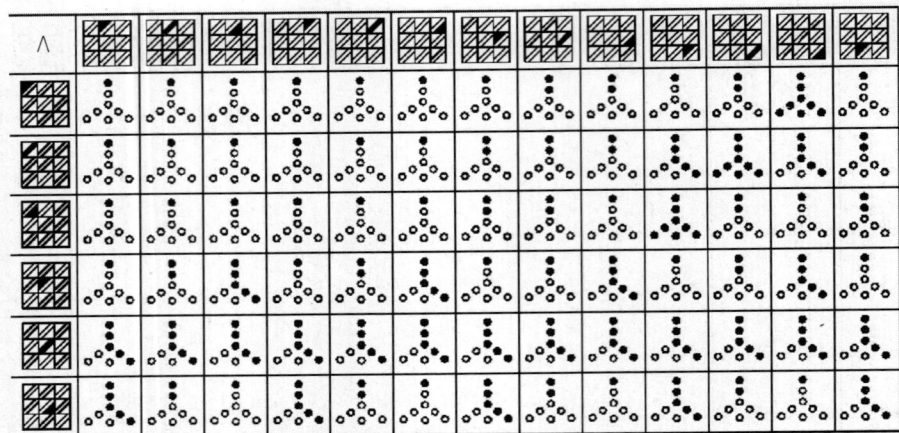

设 $i, i_1, i_2 \in \{N, NE, E, SE, S, SW, W, NW, same\}$, $j, j_1, j_2 \in \{up, between, down\}$, 那么

$$O_{i_1, up} \wedge O_{i_2, down} = O_{i_1, down} \wedge O_{i_2, up} \tag{5.81}$$

$$O_{i_1, up} \wedge O_{i_2, between} = O_{i_1, down} \wedge O_{i_2, between} \tag{5.82}$$

$$O_{i_1, up} \wedge O_{i_2, between} \rightarrow disjoint \vee meet \quad (i_1 \neq i_2) \tag{5.83}$$

$$O_{i, j_1} \wedge O_{i, j_2} \rightarrow disjoint \quad (i \neq same) \tag{5.84}$$

$$O_{i_1, j} \wedge O_{i_2, j} \rightarrow disjoint \quad (j \neq between) \tag{5.85}$$

证明:

——式(5.81)的证明如下。

因为 $\{before, meet\} \wedge \{after, metby\} \rightarrow IA$, $\{after, metby\} \wedge \{before, meet\} \rightarrow IA$, 所以

$$O_{i_1, up} \wedge O_{i_2, down} = O_{i_1, down} \wedge O_{i_2, up}$$

——式(5.82)的证明如下。

因为

$$up \wedge between = \{before, meet\} \wedge I = (\{before\} \wedge I) \cup (\{meet\} \wedge I)$$
$$\rightarrow \{before\} \cup \{before, meet\} = \{before, meet\}$$
$$\rightarrow disjoint \vee meet$$

及

$$down \wedge between = \{after, metby\} \wedge I = (\{after\} \wedge I) \cup (\{metby\} \wedge I)$$
$$\rightarrow \{after\} \cup \{after, metby\} = \{after, metby\}$$
$$\rightarrow disjoint \vee meet$$

所以

$$O_{i_1, up} \wedge O_{i_2, between} = O_{i_1, down} \wedge O_{i_2, between}$$

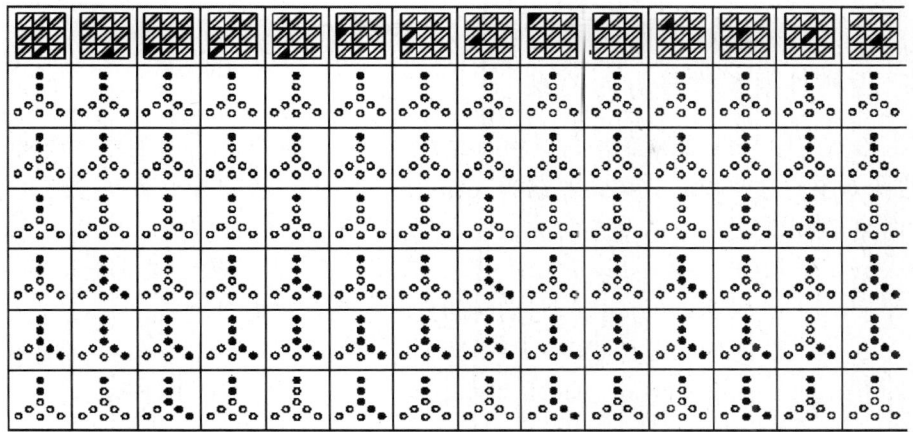

续表

——式(5.83)的证明如下。

若 $i_1 \neq i_2$，根据式(5.82)的证明以及式(5.67)和式(5.68)可证 $O_{i_1,\text{up}} \wedge O_{i_2,\text{between}} \rightarrow \text{disjoint} \vee \text{meet}$。

——式(5.84)的证明如下。

因为

$$\{\text{before}, \text{meet}\} \wedge \{\text{before}, \text{meet}\} \rightarrow \text{before}$$

$$\{\text{after}, \text{metby}\} \wedge \{\text{after}, \text{metby}\} \rightarrow \text{after}$$

所以

$$i \wedge i \quad (i \neq \text{same})$$

仅有以下 8 种推理结果

$$i \wedge i \rightarrow \{\text{before}\} \times \{\text{before}\}$$
$$i \wedge i \rightarrow \{\text{before}\} \times I$$
$$i \wedge i \rightarrow \{\text{before}\} \times \{\text{after}\}$$
$$i \wedge i \rightarrow \{\text{after}\} \times \{\text{before}\}$$
$$i \wedge i \rightarrow \{\text{after}\} \times I$$
$$i \wedge i \rightarrow \{\text{after}\} \times \{\text{after}\}$$
$$i \wedge i \rightarrow I \times \{\text{before}\}$$
$$i \wedge i \rightarrow \{\text{after}\} \times \{\text{after}\}$$

根据式(5.67)，可得 $O_{i,j_1} \wedge O_{i,j_2} \rightarrow \text{disjoint}(i \neq \text{same})$。

——式(5.85)的证明如下。

因为

$$j \wedge j \rightarrow \{\text{before}, \text{meet}\} \wedge \{\text{before}, \text{meet}\} \rightarrow \{\text{before}\},$$
$$j \wedge j \rightarrow \{\text{after}, \text{metby}\} \wedge \{\text{after}, \text{metby}\} \rightarrow \{\text{after}\}$$

所以,根据式(5.67)得

$$O_{i_1,j} \wedge O_{i_2,j} \to \text{disjoint} \quad (j \neq \text{between})$$

5.2.3 由拓扑关系推导方向关系

在三维空间中,当目标对象与参照物间的方向关系为非 $O_{\text{same,between}}$ 的单方向关系时,目标对象与参照物间的拓扑关系为 disjoint 或 meet。当目标对象与参照物间的方向关系为 $O_{\text{same,between}}$ 的单方向关系时,目标对象与参照物间的拓扑关系为 overlap,cover,contain,coveredby,inside 或 equal。当目标对象 B 与参照物 A 间的拓扑关系 $top(A,B)$ 为 contain 时,它们间的方向关系只能是 $O_{\text{same,between}}$。

定义 5.1 设映射 $TR: TOP \to P(DIR)$,$TOP = \{\text{disjoint}, \text{meet}, \text{overlap}, \text{cover}, \text{contain}, \text{coveredby}, \text{inside}, \text{equal}\}$,$DIR = \{O_{i,j} \mid i \in \{\text{N}, \text{NE}, \text{E}, \text{SE}, \text{S}, \text{SW}, \text{W}, \text{NW}, \text{same}\}, j \in \{\text{up}, \text{between}, \text{down}\}\}$,$P(DIR) = \{U \mid U \subseteq DIR\}$,使得 $TR(\text{disjoint}) = DIR - \{O_{\text{same,between}}\}$,$P(\text{meet}) = DIR - \{O_{\text{same,between}}\}$,$TR(toprel) = \{O_{\text{same,between}}\}$,$toprel \in \{\text{overlap}, \text{cover}, \text{contain}, \text{coveredby}, \text{inside}, \text{equal}\}$。

根据定义5.1,可以进行混合空间关系定性推理 $top(A,B) \wedge top(B,C) \to dir(A,C)$,定性推理的结果见表5.4。推理结果只是说明拓扑关系所蕴含方向关系的存在性,如 $\text{disjoint} \wedge \text{disjoint} \to \vee O_{i,j}$,$i \in \{\text{N}, \text{NE}, \text{E}, \text{SE}, \text{S}, \text{SW}, \text{W}, \text{NW}, \text{same}\}$,$j \in \{\text{up}, \text{between}, \text{down}\}$,表示 A 与 B 间的拓扑关系,以及 B 与 C 间的拓扑关系均为 disjoint 时,目标对象 B 与参照物 A 间的方向关系为 27 种方向关系中的任何一种均有可能。

表5.4 三维空间基于拓扑关系推导方向关系

当 A 与 B 间的拓扑关系为 contain，且 B 与 C 间的拓扑关系为 overlap 时，目标对象 C 与参照物 A 间的拓扑关系为 contain 或 cover 或 overlap。因此，C 与参照物 A 的方向关系为任何一个方向关系均有可能，所以 contain $\land$ overlap $\rightarrow$ $\lor O_{i,j}, i \in \{\text{N}, \text{NE}, \text{E}, \text{SE}, \text{S}, \text{SW}, \text{W}, \text{NW}, \text{same}\}, j \in \{\text{up}, \text{between}, \text{down}\}$。但是，根据定义 5.1，定性推理 contain $\land$ overlap $\rightarrow O_{\text{same, between}}$，$C$ 与参照物 A 的方向关系仅有 $O_{\text{same, between}}$ 可能存在。为此，重新定义拓扑关系与方向关系间的映射。

定义 5.2 设映射 $TR: TOP \rightarrow P(DIR)$，$TOP = \{\text{disjoint}, \text{meet}, \text{overlap}, \text{cover}, \text{contain}, \text{coveredby}, \text{inside}, \text{equal}\}$，$DIR = \{O_{i,j} \mid i \in \{\text{N}, \text{NE}, \text{E}, \text{SE}, \text{S}, \text{SW}, \text{W}, \text{NW}, \text{same}\}, j \in \{\text{up}, \text{between}, \text{down}\}\}$，$P(DIR) = \{U \mid U \subseteq DIR\}$，使得

$$TR(\text{disjoint}) = DIR - \{O_{\text{same, between}}\}$$
$$TR(\text{meet}) = DIR - \{O_{\text{same, between}}\}$$
$$TR(toprel) = DIR, toprel \in \{\text{overlap}, \text{coveredby}, \text{inside}\}$$
$$TR(toprel) = \{O_{\text{same, between}}\}, toprel \in \{\text{cover}, \text{contain}, \text{equal}\}$$

根据定义 5.2，得到根据拓扑关系推导方向关系的组合推理，见表 5.5。

表 5.5 三维空间基于拓扑关系推导方向关系

根据表 5.5 可得，当 $top(B,C)$ 为 overlap 或 coveredby 或 inside 时，由拓扑关系推导方向关系，没有提供方向关系约束，因为任何一种方向关系均有可能。当 $top(A,B)$ 为 overlap 时，由拓扑关系推导方向关系，也没有提供方向关系约束，任何一种方向关系均有可能。当 $top(A,B), top(B,C) \in \{\text{cover}, \text{contain}, \text{equal}\}$ 时，$dir(A,C)$ 只有 $O_{\text{same, between}}$。

第 6 章 综合拓扑方向关系描述及其推理

在日常生活中,经常说"在教学楼的南面有棵大树",这句话包含两个含义:大树相对于教学楼的方向关系是南,两者的拓扑关系是相离。如果只说大树位于教学楼的外面,大树的位置关系非常弱,因为教学楼的外面范围太大。但当知道了大树与教学楼的方向关系"南"时,大大提高了描述大树位置关系的能力。拓扑关系只是研究空间物体间的拓扑性质,当目标对象与参照物间的拓扑关系为 disjoint 时,目标对象在参照物的外部,由于参照物的外部范围很大,不利于确定目标对象与参照物间的位置关系。但当知道了目标对象与参照物间的方向关系时,大大提高了位置关系的描述能力。为提高拓扑关系的描述能力,本书用拓扑关系和方向关系描述目标对象与参照物之间的位置关系,简称综合拓扑方向关系。

本章的主要内容是分别在二维空间和三维空间中建立综合拓扑方向关系的描述、表达模型,并研究其推理。

§6.1 二维综合拓扑方向关系描述及其推理

6.1.1 二维综合拓扑方向关系描述

当目标对象与参照物之间的方向关系为非 same 时,目标对象与参照物间的拓扑关系只能是 disjoint 或 meet。第 5 章用 Allen 的 13 个区间关系描述方向区域,如用{startedby,finishedby,contain,equal}×{before,meet}描述方向区域 N。其实,可以将方向区域分为两个部分:{startedby,finishedby,contain,equal}×{before}和{startedby,finishedby,contain,equal}×{meet}。与{startedby,finishedby,contain,equal}×{before}对应的目标对象与参照物间的拓扑关系为 disjoint,与{startedby,finishedby,contain,equal}×{meet}对应的目标对象与参照物间的拓扑关系为 meet。因此,用{startedby,finishedby,contain,equal}×{before}不仅描述了目标对象 B 与参照物 A 间的方向关系 N,而且也描述了目标对象与参照物间的拓扑关系 disjoint,简写为 dN;{startedby,finishedby,contain,equal}×{meet}不仅描述了目标对象 B 与参照物 A 间的方向关系 N,而且也描述了目标对象与参照物间的拓扑关系 meet,简写为 mN。这样就大大提高了物体之间位置关系的描述能力。

第 6 章 综合拓扑方向关系描述及其推理

定义 6.1 称用拓扑关系和方向关系共同描述物体之间位置关系的方法为拓扑方向关系综合描述法。

拓扑方向关系综合描述法采用拓扑在前、方向在后的方法表示位置关系,中间不留空格,用 $topdir(A,B)$ 表示参照物 A 与 B 间的拓扑方向关系,如 $topdir(A,B)=\text{disjointE}$,简写为 $topdir(A,B)=dE$,如图 6.1(a) 所示。$topdir(A,B)=\text{meetE}$,简写为 $topdir(A,B)=mE$,如图 6.1(b) 所示。

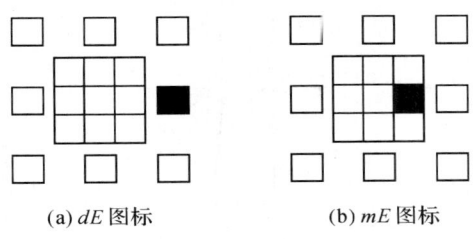

(a) dE 图标　　　(b) mE 图标

图 6.1　拓扑方向关系综合描述法图标

当物体间的拓扑关系为 overlap、cover、contain、coveredby、inside、equal,且目标对象与参照物间的方向关系为单方向关系时,两者的方向关系只能是 same。因此,为了区分两物体之间没有共同内点的拓扑关系 disjoint 和 meet,用 overlap-Nodirection 描述物体间的拓扑方向关系。根据区间关系理论,overlapNodirection 可描述为 {startedby,finishedby,contain,equal}×{startedby,finishedby,contain,equal}。

当目标对象与参照物间的拓扑方向关系为 dE 时,用 Allen 区间关系描述为 {before}×{startedby,finishedby,contain,equal}。当目标对象与参照物间的拓扑方向关系为 mE 时,用 Allen 区间关系描述为 {meet}×{startedby,finishedby,contain,equal}。其他拓扑方向关系类似,不再一一叙述。

当目标对象与参照物间的方向关系为多方向关系时,它们之间的拓扑方向关系描述仍然是拓扑在前、方向在后,如 $d(N \wedge NE)=dN \wedge dNE$。下面是用 Allen 区间关系对描述的拓扑方向关系。

$$(\text{overlap},\text{meet}) \to mN \wedge mNE \qquad (6.1)$$

$$(\text{start},\text{meet}) \to mN \wedge mNE \qquad (6.2)$$

$$(\text{during},\text{meet}) \to mNW \wedge mN \wedge mNE \qquad (6.3)$$

$$(\text{finish},\text{meet}) \to mNW \wedge mN \qquad (6.4)$$

$$(\text{overlappedby},\text{meet}) \to mNW \wedge mN \qquad (6.5)$$

$$(\text{overlap},\text{before}) \to dN \wedge dNE \qquad (6.6)$$

$$(\text{start},\text{before}) \to dN \wedge dNE \qquad (6.7)$$

$$(\text{during},\text{before}) \to dNW \wedge dN \wedge dNE \qquad (6.3)$$

$$(\text{finish}, \text{before}) \to dNW \wedge dN \tag{6.9}$$
$$(\text{overlappedby}, \text{before}) \to dNW \wedge dN \tag{6.10}$$

当将上述区间关系 before、meet 分别改为 after、metby 时，所对应的拓扑方向关系类似，不再一一叙述。

当方向关系为包含有 same 的多方向关系时，规定目标对象与参照物间的拓扑方向关系为 meet+方向关系，如图 6.2 所示。

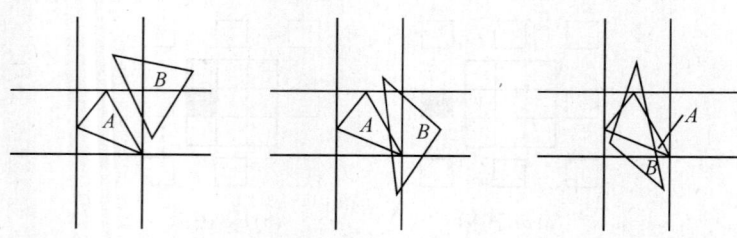

$mN \wedge \text{same} \wedge mNE \wedge mE$ $mN \wedge \text{same} \wedge mE \wedge mS \wedge mNE$ $mN \wedge \text{same} \wedge mS$

图 6.2 含有 same 的多方向关系对应的拓扑方向关系

在二维空间中，由 Allen 的区间关系对所描述的包含 same 的多方向关系对应的拓扑方向关系如下。

令 $I = \{\text{startedby}, \text{finishedby}, \text{contain}, \text{equal}\}$。

$$I \times I \to \text{same} \tag{6.11}$$
$$\{\text{overlap}\} \times I \to \text{same} \wedge mE \tag{6.12}$$
$$\{\text{start}\} \times I \to \text{same} \wedge mE \tag{6.13}$$
$$\{\text{during}\} \times I \to mW \wedge \text{same} \wedge mE \tag{6.14}$$
$$\{\text{finish}\} \times I \to mW \wedge \text{same} \tag{6.15}$$
$$\{\text{overlapped}\} \times I \to mW \wedge \text{same} \tag{6.16}$$
$$I \times \{\text{overlap}\} \to \text{same} \wedge mN \tag{6.17}$$
$$I \times \{\text{start}\} \to \text{same} \wedge mN \tag{6.18}$$
$$I \times \{\text{during}\} \to mS \wedge \text{same} \wedge mN \tag{6.19}$$
$$I \times \{\text{finish}\} \to mS \wedge \text{same} \tag{6.20}$$
$$I \times \{\text{overlappedby}\} \to mS \wedge \text{same} \tag{6.21}$$
$$\{\text{overlappedby}\} \times \{\text{overlappedby}\} \to mW \wedge \text{same} \wedge mS \wedge mSW \tag{6.22}$$
$$\{\text{overlap}\} \times \{\text{overlap}\} \to mN \wedge \text{same} \wedge mE \wedge mNE \tag{6.23}$$
$$\{\text{overlap}\} \times \{\text{overlapped}\} \to mS \wedge \text{same} \wedge mE \wedge mSE \tag{6.24}$$
$$\{\text{overlapppedby}\} \times \{\text{overlap}\} \to mW \wedge \text{same} \wedge mNW \wedge mN \tag{6.25}$$

用 Allen 区间关系可以描述方向区域（见图 6.3），方向区域 N、NE、E、SE、S、SW、W、NW、same 的定义见式(5.1)至式(5.9)。

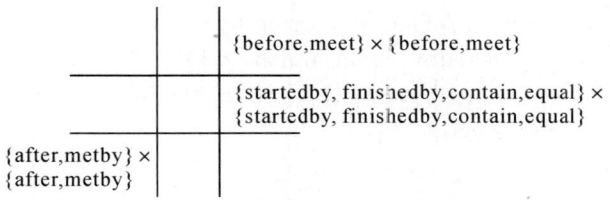

图 6.3 用 Allen 区间描述的方向关系

6.1.2 二维综合拓扑方向关系推理

综合拓扑方向关系推理是根据 A 与 B 的拓扑方向关系和 B 与 C 间的拓扑方向关系，推导 A 与 C 的拓扑方向关系。用公式描述为：$topdir(A,B) \wedge topdir(B,C) \rightarrow topdir(A,C)$。根据拓扑关系的不同，拓扑方向关系定性推理可分为四类

$$ddir(A,B) \wedge ddir(B,C) \rightarrow topdir(A,C) \quad (6.26)$$
$$ddir(A,B) \wedge mdir(B,C) \rightarrow topdir(A,C) \quad (6.27)$$
$$mdir(A,B) \wedge ddir(B,C) \rightarrow topdir(A,C) \quad (6.28)$$
$$mdir(A,B) \wedge mdir(B,C) \rightarrow topdir(A,C) \quad (6.29)$$

下面是拓扑方向关系定性推理举例。

例 6.1

$mW \wedge mS = \{\text{metby}\} \times I \wedge I \times \{\text{metby}\}$
　　　$\rightarrow \{\text{metby}\} \wedge I \times I \wedge \{\text{metby}\}$
　　　$\rightarrow \{\text{after, metby}\} \times \{\text{contain, startedby, overlappedby, metby}\}$
　　　$\rightarrow dW \vee mW \vee dSW \vee mSW$

用表 6.1 表示上述推理。

表 6.1 $mW \wedge mS$ 推理

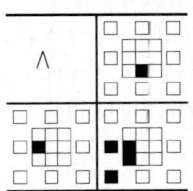

例 6.2

$mSE \wedge mW = (\text{meet, metby}) \wedge \{\text{metby}\} \times I$
　　　$\rightarrow \{\text{meet}\} \wedge \{\text{metby}\} \times \{\text{metby}\} \wedge I$
　　　$\rightarrow \{\text{finishedby, equal, finish}\} \times \{\text{after, metby}\}$
　　　$\rightarrow \{\text{finishedby, equal}\} \times \{\text{after, metby}\} \cup \{\text{finish}\} \times \{\text{after, metby}\}$
　　　$\rightarrow dS \vee mS \vee dSW \vee mSW$

例 6.3

$mSE \wedge mSW = (\text{meet, metby}) \wedge (\text{metby, metby})$

$$\rightarrow \{\text{meet}\} \wedge \{\text{metby}\} \times \{\text{metby}\} \wedge \{\text{metby}\}$$
$$\rightarrow \{\text{finishedby}, \text{equal}, \text{finish}\} \times \{\text{after}\}$$
$$\rightarrow \{\text{finishedby}, \text{equal}\} \times \{\text{after}\} \cup \{\text{finish}\} \times \{\text{after}\}$$
$$\rightarrow dS \vee dSW$$

例 6.4

$dN \wedge dS = I \times \{\text{before}\} \wedge I \times \{\text{after}\} = I \wedge I \times \{\text{before}\} \wedge \{\text{after}\}$
$\rightarrow I \times \{\text{before}, \text{meet}, \text{overlap}, \text{finishedby}, \text{contain}, \text{startedby}, \text{equal}, \text{start},$
$\quad \text{during}, \text{finish}, \text{overlappedby}, \text{metby}, \text{after}\}$
$\rightarrow I \times \{\text{before}, \text{meet}\} \cup I \times \{\text{overlap}\} \cup I \times \{\text{finishedby}, \text{contain},$
$\quad \text{startedby}, \text{equal}\} \cup I \times \{\text{start}\} \cup I \times \{\text{during}\} \cup I \times \{\text{finish}\} \cup I \times$
$\quad \{\text{overlappedby}\} \cup I \times \{\text{metby}, \text{after}\}$
$\rightarrow dN \vee mN \vee \text{same} \vee mS \vee dS$

用表 6.2 表示 $dN \wedge dS$ 推理。

表 6.2 $dN \wedge dS$ 推理

例 6.5

$dN \wedge mS = I \times \{\text{before}\} \wedge I \times \{\text{metby}\} = I \wedge I \times \{\text{before}\} \wedge \{\text{metby}\}$
$\rightarrow I \times \{\text{before}, \text{meet}, \text{overlap}, \text{start}, \text{during}\}$
$\rightarrow I \times \{\text{before}, \text{meet}\} \cup I \times \{\text{overlap}\} \cup I \times \{\text{start}\} \cup I \times \{\text{during}\}$
$\rightarrow dN \vee mN \vee \text{same} \vee mS$

例 6.6

$mN \wedge dS = I \times \{\text{meet}\} \wedge I \times \{\text{after}\} = I \wedge I \times \{\text{meet}\} \wedge \{\text{after}\}$
$\rightarrow I \times \{\text{contain}, \text{startedby}, \text{overlappedby}, \text{metby}, \text{after}\}$
$\rightarrow I \times \{\text{contain}, \text{startedby}\} \cup I \times \{\text{overlappedby}\} \cup I \times \{\text{metby}, \text{after}\}$
$\rightarrow \text{same} \vee mS \vee dS$

例 6.7

$mN \wedge mS = I \times \{\text{meet}\} \wedge I \times \{\text{metby}\} = I \wedge I \times \{\text{meet}\} \wedge \{\text{metby}\}$
$\rightarrow I \times \{\text{finishedby}, \text{equal}, \text{finish}\}$
$\rightarrow I \times \{\text{finishedby}, \text{equal}\} \cup I \times \{\text{finish}\}$
$\rightarrow \text{same} \vee mS$

例 6.8

$dN \wedge dE = I \times \{\text{before}\} \wedge \{\text{before}\} \times I$
$\rightarrow I \wedge \{\text{before}\} \times \{\text{before}\} \wedge I$
$\rightarrow \{\text{before}, \text{meet}, \text{overlap}, \text{finishedby}, \text{contain}\} \times \{\text{before}\}$

$\rightarrow \{before, meet\} \times \{before\} \bigcup \{overlap\} \times \{before\} \bigcup$
$\quad \{finishedby, contain\} \times \{before\}$
$\rightarrow dN \vee dNE$

例 6.9

$dN \wedge mE = I \times \{before\} \wedge \{meet\} \times I$
$\quad \rightarrow I \wedge \{meet\} \times \{before\} \wedge I$
$\quad \rightarrow \{meet, overlap, finishedby, contain\} \times \{before\}$
$\quad \rightarrow \{meet\} \times \{before\} \bigcup \{overlap\} \times \{before\} \bigcup \{finishedby, contain\} \times \{before\}$
$\quad \rightarrow dN \vee dNE$

例 6.10

$mN \wedge dE = I \times \{meet\} \wedge \{before\} \times I$
$\quad \rightarrow I \wedge \{before\} \times \{meet\} \wedge I$
$\quad \rightarrow \{before, meet, overlap, finishedby, contain\} \times \{before, meet\}$
$\quad \rightarrow \{before, meet\} \times \{before, meet\} \bigcup \{overlap\} \times \{before, meet\} \bigcup$
$\quad\quad \{finishedby, contain\} \times \{before, meet\}$
$\quad \rightarrow dN \vee mN \vee dNE \vee mNE$

例 6.11

$mN \wedge mE = I \times \{meet\} \wedge \{meet\} \times I$
$\quad \rightarrow I \wedge \{meet\} \times \{meet\} \times I$
$\quad \rightarrow \{contain, finishedby, overlap, meet\} \times \{before, meet\}$
$\quad \rightarrow \{contain, finishedby\} \times \{before, meet\} \bigcup \{overlap\} \times$
$\quad\quad \{before, meet\} \bigcup \{meet\} \times \{before, meet\}$
$\quad \rightarrow dN \vee dNE \vee mN \vee mNE$

拓扑方向关系式(6.26)至式(6.29)定性推理的结果分别见表 6.3 至表 6.6。

表 6.3 $ddir(A,B) \wedge ddir(B,C) \rightarrow topdir(A,C)$

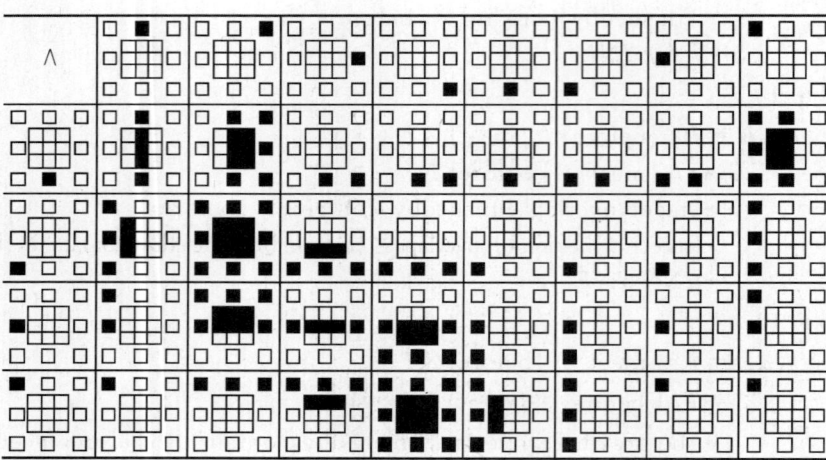

表 6.4　$ddir(A,B) \wedge mdir(B,C) \rightarrow topdir(A,C)$

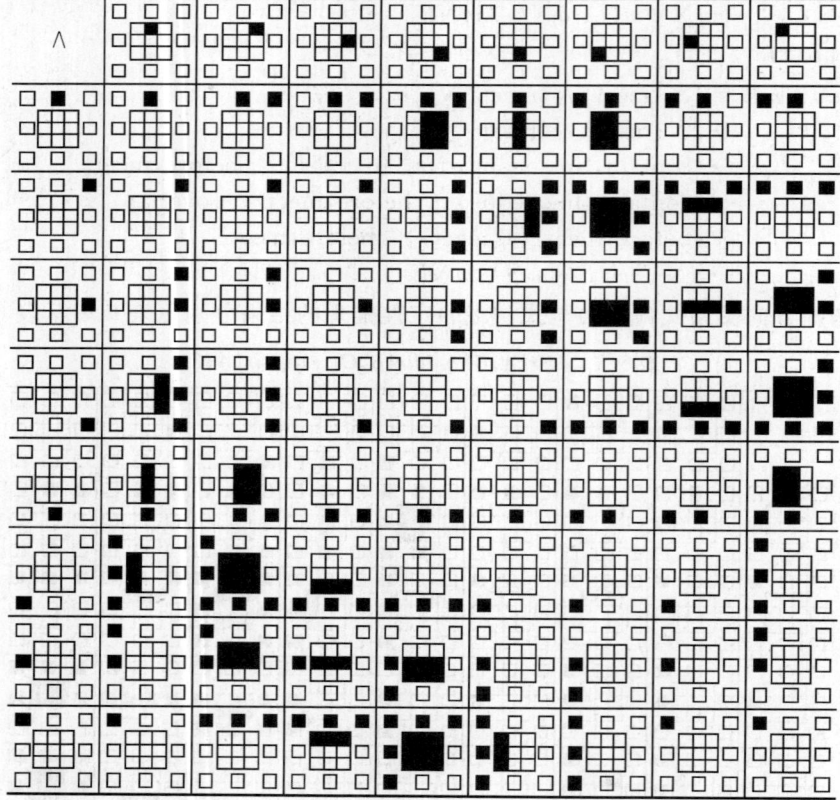

第 6 章 综合拓扑方向关系描述及其推理

表 6.5 $mdir(A,B) \wedge ddir(B,C) \rightarrow topdir(A,C)$

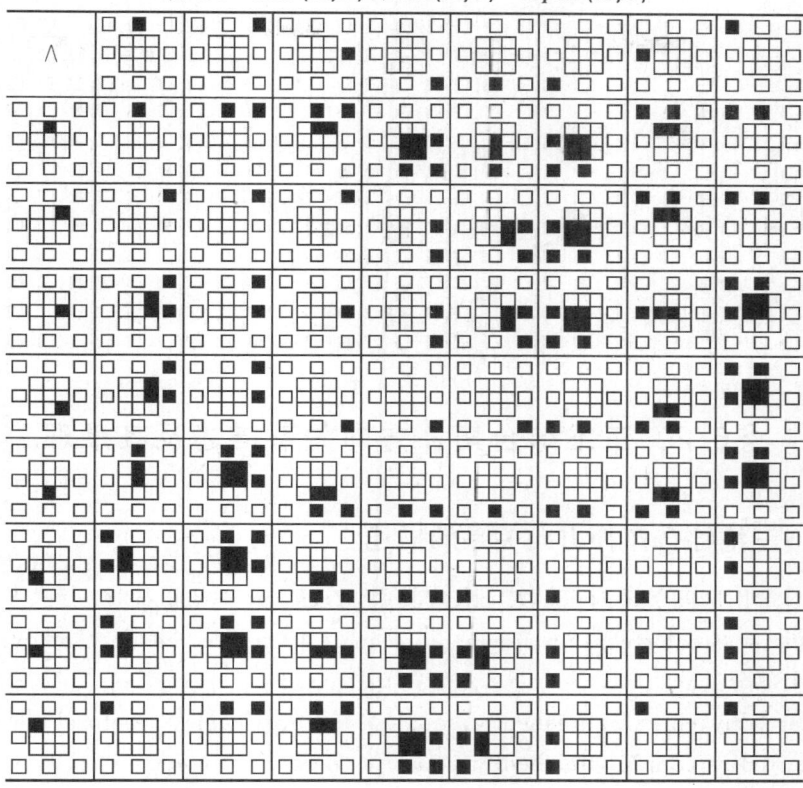

表 6.6 $mdir(A,B) \wedge mdir(B,C) \rightarrow topdir(A,C)$

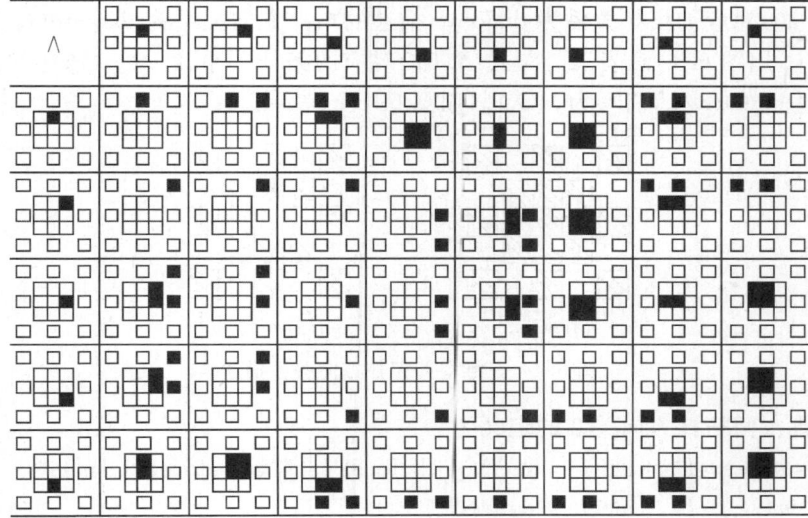

续表

§6.2 三维综合拓扑方向关系描述及其推理

6.2.1 三维综合拓扑方向关系的描述

第 5 章用区间关系对 (R_1,R_2,R_3)（其中，$R_1,R_2,R_3 \in IA$）描述三维方向区域，将 $O_{N,up}=\{startedby,contain,finishedby,equal\}\times\{before,meet\}\times\{before,meet\}$ 分为两个部分

$$\{startedby,contain,finishedby,equal\}\times\{before,meet\}\times\{before\} \bigcup$$
$$\{startedby,contain,finishedby,equal\}\times\{before\}\times\{meet\} \tag{6.30}$$

和

$$\{startedby,contain,finishedby,equal\}\times\{meet\}\times\{meet\} \tag{6.31}$$

与式(6.30)对应的拓扑关系为 disjoint，描述的拓扑方向关系为 disjoint$O_{N,up}$，简写为 $dO_{N,up}$，用图 6.4(a)表示。与式(6.31)对应的拓扑关系为 meet，描述的拓扑方向关系为 meet$O_{N,up}$，简写为 $mO_{N,up}$，用图 6.4(b)表示。

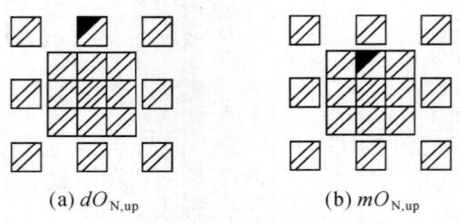

(a) $dO_{N,up}$　　　　　(b) $mO_{N,up}$

图 6.4　三维空间拓扑方向关系描述

三维空间拓扑方向关系的定义为

$$dO_{N,up}: I\times\{before,meet\}\times\{before\} \bigcup I\times\{before\}\times\{meet\} \tag{6.32}$$
$$mO_{N,up}: I\times\{meet\}\times\{meet\} \tag{6.33}$$

$$dO_{\text{N,between}}: I \times \{\text{before}\} \times I \tag{6.34}$$

$$mO_{\text{N,between}}: I \times \{\text{meet}\} \times I \tag{6.35}$$

$$dO_{\text{N,down}}: I \times \{\text{before},\text{meet}\} \times \{\text{after}\} \cup I \times \{\text{before}\} \times \{\text{metby}\} \tag{6.36}$$

$$mO_{\text{N,down}}: I \times \{\text{meet}\} \times \{\text{metby}\} \tag{6.37}$$

$$dO_{\text{NE,up}}: \{\text{before},\text{meet}\} \times \{\text{before},\text{meet}\} \times \{\text{before}\} \cup \{\text{before}\} \times$$
$$\{\text{before},\text{meet}\} \times \{\text{meet}\} \cup \{\text{meet}\} \times \{\text{before}\} \times \{\text{meet}\} \tag{6.38}$$

$$mO_{\text{NE,up}}: \{\text{meet}\} \times \{\text{meet}\} \times \{\text{meet}\} \tag{6.39}$$

$$dO_{\text{NE,between}}: \{\text{before},\text{meet}\} \times \{\text{before}\} \times I \cup \{\text{before}\} \times \{\text{meet}\} \times I \tag{6.40}$$

$$mO_{\text{NE,between}}: \{\text{meet}\} \times \{\text{meet}\} \times I \tag{6.41}$$

$$dO_{\text{NE,down}}: \{\text{before},\text{meet}\} \times \{\text{before},\text{meet}\} \times \{\text{after}\} \cup \{\text{before}\} \times$$
$$\{\text{before},\text{meet}\} \times \{\text{metby}\} \cup \{\text{meet}\} \times \{\text{before}\} \times \{\text{metby}\} \tag{6.42}$$

$$mO_{\text{NE,down}}: \{\text{meet}\} \times \{\text{meet}\} \times \{\text{metby}\} \tag{6.43}$$

$$dO_{\text{E,up}}: \{\text{before},\text{meet}\} \times I \times \{\text{before}\} \cup \{\text{before}\} \times I \times \{\text{meet}\} \tag{6.44}$$

$$mO_{\text{E,up}}: \{\text{meet}\} \times I \times \{\text{meet}\} \tag{6.45}$$

$$dO_{\text{E,between}}: \{\text{before}\} \times I \times I \tag{6.46}$$

$$mO_{\text{E,between}}: \{\text{meet}\} \times I \times I \tag{6.47}$$

$$dO_{\text{E,down}}: \{\text{before},\text{meet}\} \times I \times \{\text{after}\} \cup \{\text{before}\} \times I \times \{\text{metby}\} \tag{6.48}$$

$$mO_{\text{E,down}}: \{\text{meet}\} \times I \times \{\text{metby}\} \tag{6.49}$$

$$dO_{\text{SE,up}}: \{\text{before},\text{meet}\} \times \{\text{after},\text{metby}\} \times \{\text{before}\} \cup \{\text{before}\} \times$$
$$\{\text{after},\text{metby}\} \times \{\text{meet}\} \cup \{\text{meet}\} \times \{\text{after}\} \times \{\text{meet}\} \tag{6.50}$$

$$mO_{\text{SE,up}}: \{\text{meet}\} \times \{\text{metby}\} \times \{\text{meet}\} \tag{6.51}$$

$$dO_{\text{SE,between}}: \{\text{before},\text{meet}\} \times \{\text{after}\} \times I \cup \{\text{before}\} \times \{\text{metby}\} \times I \tag{6.52}$$

$$mO_{\text{SE,between}}: \{\text{meet}\} \times \{\text{metby}\} \times I \tag{6.53}$$

$$dO_{\text{SE,down}}: \{\text{before},\text{meet}\} \times \{\text{after},\text{metby}\} \times \{\text{after}\} \cup \{\text{before}\} \times$$
$$\{\text{after},\text{metby}\} \times \{\text{metby}\} \cup \{\text{meet}\} \times \{\text{after}\} \times \{\text{metby}\} \tag{6.54}$$

$$mO_{\text{SE,down}}: \{\text{meet}\} \times \{\text{metby}\} \times \{\text{metby}\} \tag{6.55}$$

$$dO_{\text{S,up}}: I \times \{\text{after},\text{metby}\} \times \{\text{before}\} \cup I \times \{\text{after}\} \times \{\text{meet}\} \tag{6.56}$$

$$mO_{\text{S,up}}: I \times \{\text{metby}\} \times \{\text{meet}\} \tag{6.57}$$

$$dO_{\text{S,between}}: I \times \{\text{after}\} \times I \tag{6.58}$$

$$mO_{\text{S,between}}: I \times \{\text{metby}\} \times I \tag{6.59}$$

$$dO_{\text{S,down}}: I \times \{\text{after},\text{metby}\} \times \{\text{after}\} \cup I \times \{\text{after}\} \times \{\text{metby}\} \tag{6.60}$$

$$mO_{\text{S,down}}: I \times \{\text{metby}\} \times \{\text{metby}\} \tag{6.61}$$

$$dO_{\text{SW,up}}: \{\text{after},\text{metby}\} \times \{\text{after},\text{metby}\} \times \{\text{before}\} \cup \{\text{after}\} \times$$
$$\{\text{after},\text{metby}\} \times \{\text{meet}\} \cup \{\text{metby}\} \times \{\text{after}\} \times \{\text{meet}\} \tag{6.62}$$

$$mO_{\text{SW,up}}: \{\text{metby}\} \times \{\text{metby}\} \times \{\text{meet}\} \tag{6.63}$$

$$dO_{\text{SW,between}}: \{\text{after}\} \times \{\text{after},\text{metby}\} \times I \cup \{\text{metby}\} \times \{\text{after}\} \times I \tag{6.64}$$

$mO_{\text{SW,between}} : \{\text{metby}\} \times \{\text{metby}\} \times I$ \hfill (6.65)

$dO_{\text{SW,down}} : \{\text{after, metby}\} \times \{\text{after, metby}\} \times \{\text{after}\} \bigcup \{\text{after}\} \times$
$\{\text{after, metby}\} \times \{\text{metby}\} \bigcup \{\text{metby}\} \times \{\text{after}\} \times \{\text{metby}\}$ \hfill (6.66)

$mO_{\text{SW,down}} : \{\text{metby}\} \times \{\text{metby}\} \times \{\text{metby}\}$ \hfill (6.67)

$dO_{\text{W,up}} : \{\text{after, metby}\} \times I \times \{\text{before}\} \bigcup \{\text{after}\} \times I \times \{\text{meet}\}$ \hfill (6.68)

$mO_{\text{W,up}} : \{\text{metby}\} \times I \times \{\text{meet}\}$ \hfill (6.69)

$dO_{\text{W,between}} : \{\text{after}\} \times I \times I$ \hfill (6.70)

$mO_{\text{W,between}} : \{\text{metby}\} \times I \times I$ \hfill (6.71)

$dO_{\text{W,down}} : \{\text{after, metby}\} \times I \times \{\text{after}\} \bigcup \{\text{after}\} \times I \times \{\text{metby}\}$ \hfill (6.72)

$mO_{\text{W,down}} : \{\text{metby}\} \times I \times \{\text{metby}\}$ \hfill (6.73)

$dO_{\text{NW,up}} : \{\text{after, metby}\} \times \{\text{before, meet}\} \times \{\text{before}\} \bigcup \{\text{after}\} \times$
$\{\text{before, meet}\} \times \{\text{meet}\} \bigcup \{\text{metby}\} \times \{\text{before}\} \times \{\text{meet}\}$ \hfill (6.74)

$mO_{\text{NW,up}} : \{\text{metby}\} \times \{\text{meet}\} \times \{\text{meet}\}$ \hfill (6.75)

$dO_{\text{NW,between}} : \{\text{after}\} \times \{\text{before, meet}\} \times I \bigcup \{\text{metby}\} \times \{\text{before}\} \times I$ \hfill (6.76)

$mO_{\text{NW,between}} : \{\text{metby}\} \times \{\text{meet}\} \times I$ \hfill (6.77)

$dO_{\text{NW,down}} : \{\text{after, metby}\} \times \{\text{before, meet}\} \times \{\text{after}\} \bigcup \{\text{after}\} \times$
$\{\text{before, meet}\} \times \{\text{metby}\} \bigcup \{\text{metby}\} \times \{\text{before}\} \times \{\text{metby}\}$ \hfill (6.78)

$mO_{\text{NW,down}} : \{\text{metby}\} \times \{\text{meet}\} \times \{\text{metby}\}$ \hfill (6.79)

$dO_{\text{same,up}} : I \times I \times \{\text{before}\}$ \hfill (6.80)

$mO_{\text{same,up}} : I \times I \times \{\text{meet}\}$ \hfill (6.81)

$O_{\text{same,between}} : I \times I \times I$ \hfill (6.82)

$dO_{\text{same,down}} : I \times I \times \{\text{after}\}$ \hfill (6.83)

$mO_{\text{same,down}} : I \times I \times \{\text{metby}\}$ \hfill (6.84)

用图 6.5 描述三维空间中的拓扑方向关系。

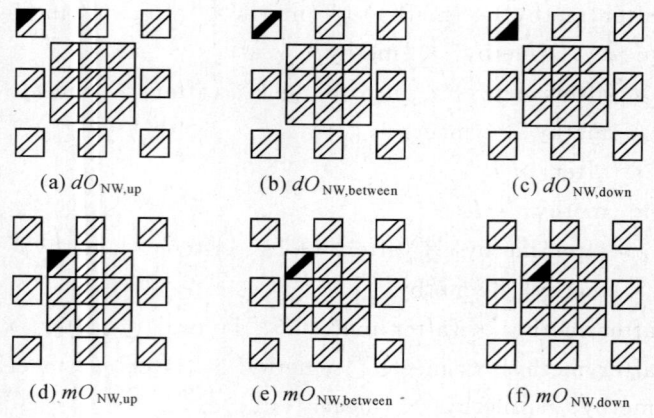

(a) $dO_{\text{NW,up}}$ (b) $dO_{\text{NW,between}}$ (c) $dO_{\text{NW,down}}$

(d) $mO_{\text{NW,up}}$ (e) $mO_{\text{NW,between}}$ (f) $mO_{\text{NW,down}}$

图 6.5 三维空间拓扑方向关系的描述图标

6.2.2 三维综合拓扑方向关系定性推理

三维空间综合拓扑方向关系定性推理是根据 A 与 B 的拓扑方向关系和 B 与 C 间的拓扑方向关系,推导 A 与 C 的拓扑方向关系。用公式描述为: $topdir(A,B) \wedge topdir(B,C) \rightarrow topdir(A,C)$。与二维空间关系类似,根据拓扑关系的不同,三维空间拓扑方向关系定性推理分为四类,见式(6.26)至式(6.29)。

1. 当拓扑关系为 $top(A,B)$ = disjoint 且 $top(B,C)$ = disjoint 时

当 A 与 B 的拓扑方向关系中的拓扑关系以及 B 与 C 间的拓扑方向关系中的拓扑关系均为 disjoint 时,可以推导 A 与 C 的拓扑方向关系,用公式描述为 $ddir(A,B) \wedge ddir(A,B) \rightarrow topdir(A,C)$。

令

IA = {before, meet, overlap, finishedby, contain, startedby, equal, start, during, finish, overlappedby, metby, after}

则

$dO_{N,up} \wedge dO_{SE,up} = (I \times \{before, meet\} \times \{before\} \vee I \times \{before\} \times \{meet\}) \wedge (\{before, meet\} \times \{after, metby\} \times \{before\} \cup \{before\} \times \{after, metby\} \times \{meet\} \cup \{meet\} \times \{after\} \times \{meet\})$

$\rightarrow \{before, meet, contain, finishedby, overlap\} \times IA \times \{before\}$

$\rightarrow dO_{N,up} \vee dO_{NE,up} \vee dO_{same,up} \vee dO_{E,up} \vee dO_{S,up} \vee dO_{SE,up}$

$dO_{N,up} \wedge dO_{SE,between} = (I \times \{before, meet\} \times \{before\} \vee I \times \{before\} \times \{meet\}) \wedge (\{before, meet\} \times \{after\} \times I \cup \{before\} \times \{metby\} \times I)$

$\rightarrow \{before, meet, contain, finishedby, overlap\} \times IA \times \{before, meet\}$

$\rightarrow dO_{N,up} \vee dO_{NE,up} \vee dO_{same,up} \vee dO_{E,up} \vee dO_{S,up} \vee dO_{SE,up} \vee mO_{N,up} \vee mO_{NE,up} \vee mO_{same,up} \vee mO_{E,up} \vee mO_{S,up} \vee mO_{SE,up}$

$dO_{N,up} \wedge dO_{SE,down} = (I \times \{before, meet\} \times \{before\} \vee I \times \{before\} \times \{meet\}) \wedge (\{before, meet\} \times \{after, metby\} \times \{after\} \cup \{before\} \times \{after, metby\} \times \{after, metby\} \cup \{meet\} \times \{after\} \times \{after\})$

$\rightarrow \{\text{before}, \text{meet}, \text{contain}, \text{finishedby}, \text{overlap}\} \times IA \times IA$

$\rightarrow dO_{\text{N},i} \vee dO_{\text{NE},i} \vee dO_{\text{E},i} \vee dO_{\text{same},i} \vee dO_{\text{S},i} \vee dO_{\text{SE},i} \vee mO_{\text{N},i} \vee mO_{\text{NE},i} \vee O_{\text{same},i} \vee mO_{\text{E},i} \vee mO_{\text{S},i} \vee mO_{\text{SE},i}$

式中,$i \in \{\text{up}, \text{between}, \text{down}\}$。

$dO_{\text{N},\text{between}} \wedge dO_{\text{SE},\text{up}} = I \times \{\text{before}\} \times I \wedge (\{\text{before}, \text{meet}\} \times \{\text{after}, \text{metby}\} \times \{\text{before}\} \cup \{\text{before}\} \times \{\text{after}, \text{metby}\} \times \{\text{meet}\} \cup \{\text{meet}\} \times \{\text{after}\} \times \{\text{meet}\})$

$\rightarrow \{\text{before}, \text{meet}, \text{contain}, \text{finishedby}, \text{overlap}\} \times IA \times \{\text{before}, \text{meet}, \text{contain}, \text{finishedby}, \text{overlap}\}$

$\rightarrow dO_{\text{N},i} \vee dO_{\text{NE},i} \vee dO_{\text{E},i} \vee dO_{\text{same},i} \vee dO_{\text{S},i} \vee dO_{\text{SE},i} \vee mO_{\text{N},i} \vee mO_{\text{NE},i} \vee O_{\text{same},i} \vee mO_{\text{E},i} \vee mO_{\text{S},i} \vee mO_{\text{SE},i}$

式中,$i \in \{\text{up}, \text{between}\}$。

$dO_{\text{N},\text{between}} \wedge dO_{\text{SE},\text{between}} = I \times \{\text{before}\} \times I \wedge (\{\text{before}, \text{meet}\} \times \{\text{after}\} \times I \cup \{\text{before}\} \times \{\text{metby}\} \times I)$

$\rightarrow \{\text{before}, \text{meet}, \text{contain}, \text{finishedby}, \text{overlap}\} \times IA \times I$

$\rightarrow dO_{\text{N},\text{between}} \vee dO_{\text{NE},\text{between}} \vee dO_{\text{E},\text{between}} \vee dO_{\text{S},\text{between}} \vee dO_{\text{SE},\text{between}} \vee mO_{\text{N},\text{between}} \vee mO_{\text{NE},\text{between}} \vee O_{\text{same},\text{between}} \vee mO_{\text{E},\text{between}} \vee mO_{\text{S},\text{between}} \vee mO_{\text{SE},\text{between}}$

$dO_{\text{N},\text{between}} \wedge dO_{\text{SE},\text{down}} = I \times \{\text{before}\} \times I \wedge (\{\text{before}, \text{meet}\} \times \{\text{after}, \text{metby}\} \times \{\text{after}\} \cup \{\text{before}\} \times \{\text{after}, \text{metby}\} \times \{\text{after}, \text{metby}\} \cup \{\text{meet}\} \times \{\text{after}\} \times \{\text{after}\})$

$\rightarrow \{\text{before}, \text{meet}, \text{contain}, \text{finishedby}, \text{overlap}\} \times IA \times \{\text{after}, \text{metby}, \text{overlappedby}, \text{startedby}, \text{contain}\}$

$\rightarrow dO_{\text{N},i} \vee dO_{\text{NE},i} \vee dO_{\text{E},i} \vee dO_{\text{same},i} \vee dO_{\text{S},i} \vee dO_{\text{SE},i} \vee mO_{\text{N},i} \vee mO_{\text{NE},i} \vee O_{\text{same},i} \vee mO_{\text{E},i} \vee mO_{\text{S},i} \vee mO_{\text{SE},i}$

式中,$i \in \{\text{between}, \text{down}\}$。

$dO_{\text{N},\text{down}} \wedge dO_{\text{SE},\text{up}} = (I \times \{\text{before}, \text{meet}\} \times \{\text{after}\} \vee I \times \{\text{before}\} \times \{\text{metby}\}) \wedge (\{\text{before}, \text{meet}\} \times \{\text{after}, \text{metby}\} \times \{\text{before}\} \cup \{\text{before}\} \times \{\text{after}, \text{metby}\} \times \{\text{meet}\} \cup \{\text{meet}\} \times \{\text{after}\} \times \{\text{meet}\})$

$\rightarrow \{\text{before}, \text{meet}, \text{contain}, \text{finishedby}, \text{overlap}\} \times IA \times IA$

$\rightarrow dO_{\text{N},i} \vee dO_{\text{NE},i} \vee dO_{\text{E},i} \vee dO_{\text{same},i} \vee dO_{\text{S},i} \vee dO_{\text{SE},i} \vee mO_{\text{N},i} \vee mO_{\text{NE},i} \vee O_{\text{same},i} \vee mO_{\text{E},i} \vee mO_{\text{S},i} \vee mO_{\text{SE},i}$

式中, $i \in \{\text{up}, \text{between}, \text{down}\}$。

$dO_{\text{N,down}} \wedge dO_{\text{SE,between}} = (I \times \{\text{before}, \text{meet}\} \times \{\text{after}\} \vee I \times \{\text{before}\} \times \{\text{metby}\}) \wedge$
$\quad (\{\text{before}, \text{meet}\} \times \{\text{after}\} \times I \cup \{\text{before}\} \times \{\text{metby}\} \times I)$
$\quad \rightarrow \{\text{before}, \text{meet}, \text{contain}, \text{finishedby}, \text{overlap}\} \times IA \times$
$\quad \{\text{after}, \text{meet}\}$
$\quad \rightarrow dO_{\text{N,down}} \vee dO_{\text{NE,down}} \vee dO_{\text{same,down}} \vee dO_{\text{E,down}} \vee dO_{\text{S,down}} \vee$
$\quad dO_{\text{SE,down}} \vee mO_{\text{N,down}} \vee mO_{\text{NE,down}} \vee mO_{\text{same,down}} \vee mO_{\text{E,down}} \vee$
$\quad mO_{\text{S,down}} \vee mO_{\text{SE,down}}$

$dO_{\text{N,down}} \wedge dO_{\text{SE,down}} = (I \times \{\text{before}, \text{meet}\} \times \{\text{after}\} \vee I \times \{\text{before}\} \times$
$\quad \{\text{metby}\}) \wedge (\{\text{before}, \text{meet}\} \times \{\text{after}, \text{metby}\} \times$
$\quad \{\text{after}\} \cup \{\text{before}\} \times \{\text{after}, \text{metby}\} \times \{\text{after}, \text{metby}\} \cup$
$\quad \{\text{meet}\} \times \{\text{after}\} \times \{\text{after}\})$
$\quad \rightarrow \{\text{before}, \text{meet}, \text{contain}, \text{finishedby}, \text{overlap}\} \times IA \times \{\text{after}\}$
$\quad \rightarrow dO_{\text{N,down}} \vee dO_{\text{NE,down}} \vee dO_{\text{same,down}} \vee dO_{\text{E,down}} \vee$
$\quad dO_{\text{S,down}} \vee dO_{\text{SE,down}}$

令 $DIRXY = \{\text{N}, \text{NE}, \text{E}, \text{SE}, \text{S}, \text{SW}, \text{N}, \text{NW}, \text{same}\}$, $DIRZ = \{\text{up}, \text{between}, \text{down}\}$。上面 9 个关系的推理结果用表 6.7 表示,表中 $j_1, j_2 \in DIRZ$。

表 6.7 $dO_{\text{N},j_1} \wedge dO_{\text{SE},j_2} \rightarrow topdir(A, C)$

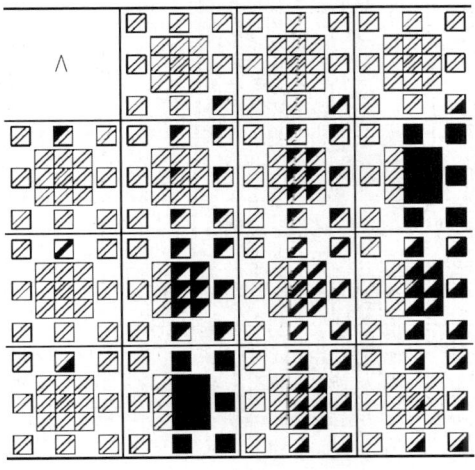

定性推理 $dO_{i_1,j_1} \wedge dO_{i_2,j_2} \rightarrow topdir(A, C)$ 的结果见表 6.8。表中 $i_1, i_2 \in DIRXY$, $j_1, j_2 \in DIRZ$。

表 6.8 $dO_{i_1,j_1} \wedge dO_{i_2,j_2}$

→$topdir(A,C)$

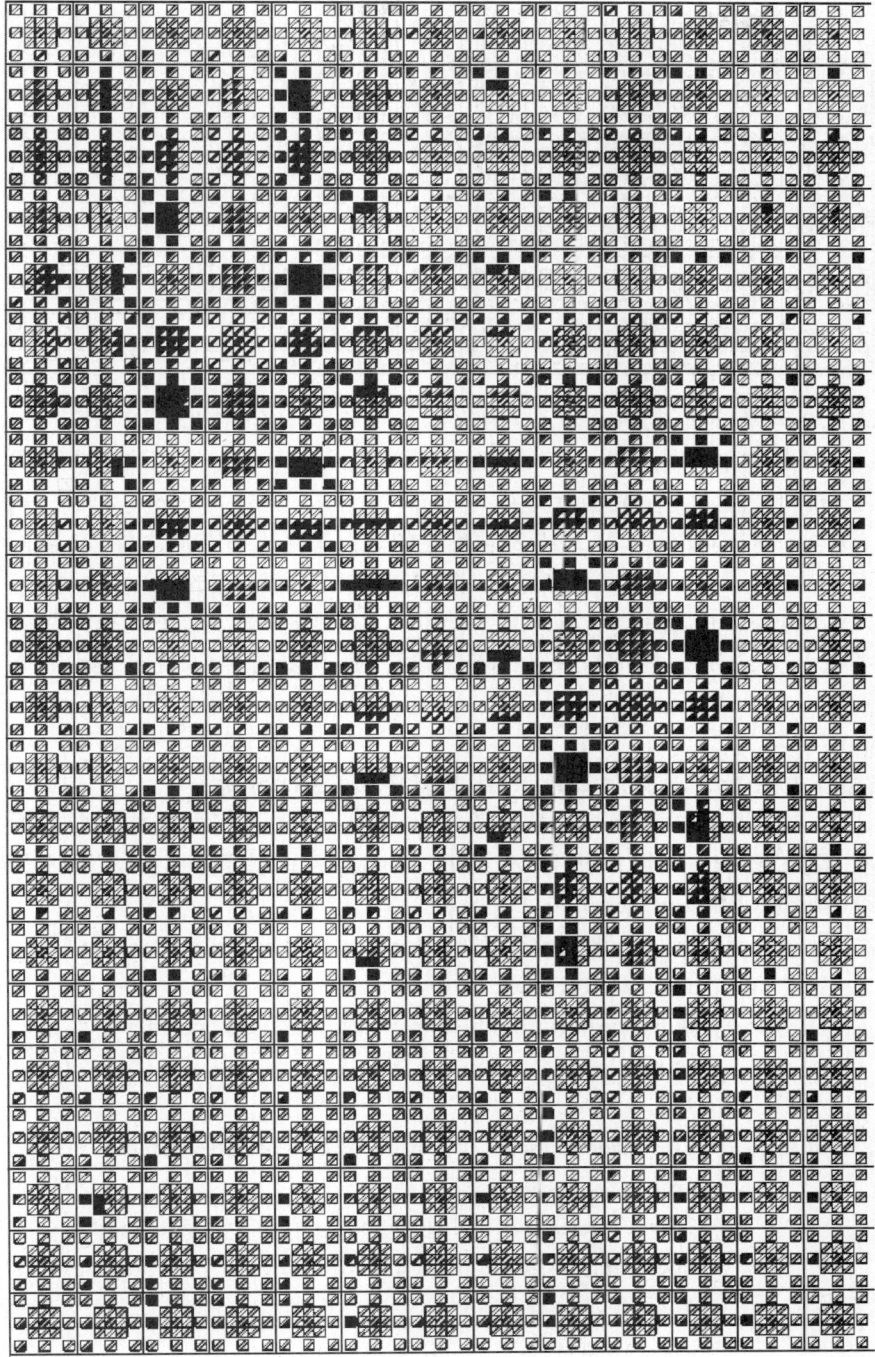

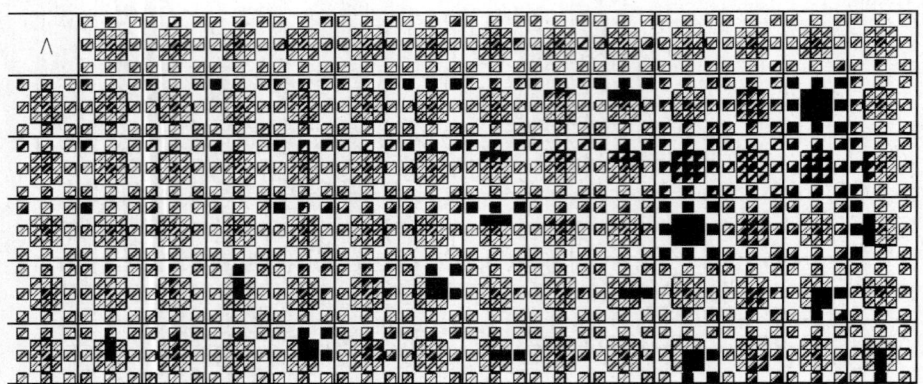

根据表 6.8 可得，对给定的 $i_1,i_2 \in DIRXY, j_1,j_2 \in DIRZ$，由 $dO_{i_1,j_1} \wedge dO_{i_2,j_2} \to topdir(A,C)$ 得出的方向关系 $dir(A,C)$ 与 $O_{i_1,j_1} \wedge O_{i_2,j_2} \to dir(A,C)$ 得出的方向关系 $dir(A,C)$ 相同。当 $i_1=i_2$ 或 $j_1=j_2$ 时，A 与 C 的拓扑关系为 disjoint，而且，只要 $dir(A,B)$ 与 $dir(B,C)$ 在由 X、Y 或 Z 轴确定的方向关系上有共同之处，A 与 C 的拓扑关系就为 disjoint，如 $O_{N,up}$ 和 $O_{NE,down}$ 分别属于由 Y 轴确定的 $\{(x,y,z)|y > \max\{y_A\},(x_A,y_A,z_A) \in A\}$ 和 $\{(x,y,z)|y > \max\{y_B\},(x_B,y_B,z_B) \in B\}$ 时，A 与 C 的拓扑关系为 disjoint。当 $i_1,i_2 \in \{N,E,S,W\}$ 时，只要 $i_1 \neq i_2$，就有 $dO_{i_1,up} \wedge dO_{i_2,down} = dO_{i_1,down} \wedge dO_{i_2,up}$，且 $top(A,C) =$ disjoint $\vee$ meet，其他情况时，$top(A,C) =$ disjoint。

2. 当拓扑关系 $top(A,B) =$ disjoint 且 $top(B,C) =$ meet 时

当 A 与 B 的拓扑方向关系中的拓扑关系为 disjoint、B 与 C 的拓扑方向关系中的拓扑关系为 meet 时，可以推导 A 与 C 的拓扑方向关系，用公式描述为 $ddir(A,B) \wedge mdir(B,C) \to topdir(A,C)$。

$dO_{N,up} \wedge mO_{E,up} = (I \times \{before, meet\} \times \{before\} \vee I \times \{before\} \times \{meet\}) \wedge \{meet\} \times I \times \{meet\}$
$\to \{meet, contain, finishedby, overlap\} \times \{before, meet\} \times \{before\}$
$\to dO_{N,up} \vee dO_{NE,up}$

$dO_{N,up} \wedge mO_{E,between} = (I \times \{before, meet\} \times \{before\} \vee I \times \{before\} \times \{meet\}) \wedge \{meet\} \times I \times I$
$\to \{meet, contain, finishedby, overlap\} \times \{before, meet\} \times \{before\} \cup \{meet, contain, finishedby, overlap\} \times \{before\} \times \{before, meet\}$
$\to dO_{N,up} \vee dO_{NE,up}$

第 6 章 综合拓扑方向关系描述及其推理　　　　　　　　　　　　　129

续表

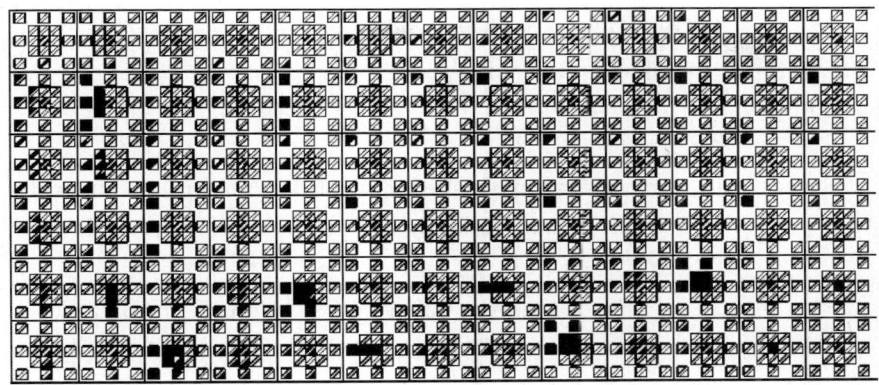

$$dO_{N,up} \wedge mO_{E,down} = (I \times \{before,meet\} \times \{before\} \vee I \times \{before\} \times \{meet\}) \wedge$$
$$\{meet\} \times I \times \{metby\}$$
$$\rightarrow \{meet,contain,finishedby,overlap\} \times \{before,meet\} \times \{before,$$
$$meet,overlap,start,during\} \cup \{meet,contain,finishedby,$$
$$overlap\} \times \{before\} \times \{finishedby,equal,overlappedby\}$$
$$\rightarrow dO_{N,j} \vee dO_{NE,j} \vee mO_{N,j} \vee mO_{NE,j}$$

式中，$j \in \{up,between,down\}$。

$$dO_{N,between} \wedge mO_{E,up} = I \times \{before\} \times I \wedge \{meet\} \times I \times \{meet\}$$
$$\rightarrow \{meet,contain,finishedby,overlap\} \times \{before\} \times$$
$$\{meet,contain,finishedby,overlap\}$$
$$\rightarrow dO_{N,j} \vee dO_{NE,j}$$

式中，$j \in \{up,between\}$。

$$dO_{N,between} \wedge mO_{E,between} = I \times \{before\} \times I \wedge \{meet\} \times I \times I$$
$$\rightarrow \{meet,contain,finishedby,overlap\} \times \{before\} \times I$$
$$\rightarrow dO_{N,between} \vee dO_{NE,between}$$

$$dO_{N,between} \wedge mO_{E,down} = I \times \{before\} \times I \wedge \{meet\} \times I \times \{metby\}$$
$$\rightarrow \{meet,contain,finishedby,overlap\} \times \{before\} \times$$
$$\{contain,startedby,overlappedby,metby\}$$
$$\rightarrow dO_{N,j} \vee dO_{NE,j}$$

式中，$j \in \{between,down\}$。

$$dO_{N,down} \wedge mO_{E,up} = (I \times \{before,meet\} \times \{after\} \vee I \times \{before\} \times \{metby\}) \wedge$$
$$\{meet\} \times I \times \{meet\}$$
$$\rightarrow \{meet,contain,finishedby,overlap\} \times \{before,meet\} \times \{after,$$
$$metby,overlappedby,finish,during\} \cup \{meet,contain,$$

$$\text{finishedby, overlap}\} \times \{\text{before}\} \times \{\text{startededby, equal, start}\}$$
$$\rightarrow dO_{\text{N},j} \vee dO_{\text{NE},j} \vee mO_{\text{N},j} \vee mO_{\text{NE},j}$$

式中, $j \in \{\text{up, between, down}\}$。

$$dO_{\text{N,down}} \wedge mO_{\text{E,between}} = (I \times \{\text{before, meet}\} \times \{\text{after}\} \vee I \times \{\text{before}\} \times \{\text{metby}\}) \wedge$$
$$\{\text{meet}\} \times I \times I$$
$$\rightarrow \{\text{meet, contain, finishedby, overlap}\} \times \{\text{before}\} \times \{\text{after}\} \cup$$
$$\{\text{meet, contain, finishedby, overlap}\} \times \{\text{before}\} \times \{\text{after, metby}\}$$
$$\rightarrow dO_{\text{N,down}} \vee dO_{\text{NE,down}}$$

$$dO_{\text{N,down}} \wedge mO_{\text{E,down}} = (I \times \{\text{before, meet}\} \times \{\text{after}\} \vee I \times \{\text{before}\} \times \{\text{metby}\}) \wedge$$
$$\{\text{meet}\} \times I \times \{\text{metby}\}$$
$$\rightarrow \{\text{meet, contain, finishedby, overlap}\} \times \{\text{before, meet}\} \times \{\text{after}\}$$
$$\rightarrow dO_{\text{N,down}} \vee dO_{\text{NE,down}}$$

以上 9 种关系的推理结果用表 6.9 表示, 表中 $j_1, j_2 \in DIRZ$。

表 6.9 $dO_{\text{N},j_1} \wedge mO_{\text{E},j_2} \rightarrow topdir(A,C)$

$$dO_{\text{N,up}} \wedge mO_{\text{SE,up}} = (I \times \{\text{before, meet}\} \times \{\text{before}\} \cup I \times \{\text{before}\} \times \{\text{meet}\}) \wedge \{\text{meet}\}$$
$$\{\text{metby}\} \times \{\text{meet}\}$$
$$\rightarrow \{\text{meet, contain, finishedby, overlap}\} \times \{\text{before, meet, overlap,}$$
$$\text{start, during, finishedby, equal, finish}\} \times \{\text{before}\}$$
$$\rightarrow dO_{\text{N,up}} \vee dO_{\text{NE,up}} \vee dO_{\text{same,up}} \vee dO_{\text{E,up}} \vee dO_{\text{S,up}} \vee dO_{\text{SE,up}}$$

$$dO_{\text{N,up}} \wedge mO_{\text{SE,between}} = (I \times \{\text{before, meet}\} \times \{\text{before}\} \cup I \times \{\text{before}\} \times \{\text{meet}\}) \wedge$$
$$\{\text{meet}\} \times \{\text{metby}\} \times I$$
$$\rightarrow \{\text{meet, contain, finishedby, overlap}\} \times \{\text{before, meet,}$$
$$\text{overlap, start, during, finishedby, equal, finish}\} \times$$

$$\{\text{before}\} \bigcup \{\text{meet}, \text{contain}, \text{finishedby}, \text{overlap}\} \times$$
$$\{\text{before}, \text{meet}, \text{overlap}, \text{start}, \text{during}\} \times \{\text{before}, \text{meet}\}$$
$$\rightarrow dO_{\text{N,up}} \vee dO_{\text{NE,up}} \vee dO_{\text{same,up}} \vee dO_{\text{E,up}} \vee dO_{\text{S,up}} \vee dO_{\text{SE,up}} \vee$$
$$mO_{\text{N,up}} \vee mO_{\text{NE,up}} \vee mO_{\text{same,up}} \vee mO_{\text{E,up}} \vee mO_{\text{S,up}} \vee mO_{\text{SE,up}}$$

$$dO_{\text{N,up}} \wedge mO_{\text{SE,down}} = (I \times \{\text{before}, \text{meet}\} \times \{\text{before}\} \bigcup I \times \{\text{before}\} \times \{\text{meet}\}) \wedge$$
$$\{\text{meet}\} \times \{\text{metby}\} \times \{\text{metby}\}$$
$$\rightarrow \{\text{meet}, \text{contain}, \text{finishedby}, \text{overlap}\} \times \{\text{before}, \text{meet}, \text{overlap},$$
$$\text{start}, \text{during}, \text{finishedby}, \text{equal}, \text{finish}\} \times \{\text{before}, \text{meet}, \text{overlap},$$
$$\{\text{start}, \text{during}\} \bigcup \{\text{meet}, \text{contain}, \text{finishedby}, \text{overlap}\} \times \{\text{before},$$
$$\text{meet}, \text{overlap}, \text{start}, \text{during}\} \times \{\text{finishedby}, \text{equal}, \text{finish}\}$$
$$\rightarrow dO_{\text{N,up}} \vee dO_{\text{NE,up}} \vee dO_{\text{same,up}} \vee dO_{\text{E,up}} \vee dO_{\text{S,up}} \vee dO_{\text{SE,up}} \vee$$
$$dO_{\text{N,between}} \vee dO_{\text{NE,between}} \vee dO_{\text{N,down}} \vee dO_{\text{NE,down}} \vee mO_{\text{N,up}} \vee$$
$$mO_{\text{NE,up}} \vee mO_{\text{same,up}} \vee mO_{\text{E,up}} \vee mO_{\text{S,up}} \vee mO_{\text{SE,up}} \vee$$
$$mO_{\text{N,between}} \vee mO_{\text{NE,between}} \vee O_{\text{same,between}} \vee mO_{\text{E,between}} \vee$$
$$mO_{\text{S,between}} \vee mO_{\text{SE,between}} \vee mO_{\text{N,down}} \vee mO_{\text{NE,down}} \vee$$
$$mO_{\text{same,down}} \vee mO_{\text{E,down}} \vee mO_{\text{S,down}} \vee mO_{\text{SE,down}}$$

$$dO_{\text{N,between}} \wedge mO_{\text{SE,up}} = I \times \{\text{before}\} \times I \wedge \{\text{meet}\} \times \{\text{metby}\} \times \{\text{meet}\}$$
$$\rightarrow \{\text{meet}, \text{contain}, \text{finishedby}, \text{overlap}\} \times \{\text{before}, \text{meet}, \text{overlap},$$
$$\text{start}, \text{during}\} \times \{\text{meet}, \text{contain}, \text{finishedby}, \text{overlap}\}$$
$$\rightarrow dO_{\text{N,up}} \vee dO_{\text{NE,up}} \vee dO_{\text{N,between}} \vee dO_{\text{NE,between}} \vee mO_{\text{N,up}} \vee mO_{\text{NE,up}} \vee$$
$$mO_{\text{same,up}} \vee mO_{\text{E,up}} \vee mO_{\text{S,up}} \vee mO_{\text{SE,up}} \vee mO_{\text{N,between}} \vee mO_{\text{NE,between}} \vee$$
$$O_{\text{same,between}} \vee mO_{\text{E,between}} \vee mO_{\text{S,between}} \vee mO_{\text{SE,between}}$$

$$dO_{\text{N,between}} \wedge mO_{\text{SE,between}} = I \times \{\text{before}\} \times I \wedge \{\text{meet}\} \times \{\text{metby}\} \times I$$
$$\rightarrow \{\text{meet}, \text{contain}, \text{finishedby}, \text{overlap}\} \times$$
$$\{\text{before}, \text{meet}, \text{overlap}, \text{start}, \text{during}\} \times I$$
$$\rightarrow dO_{\text{N,between}} \vee dO_{\text{NE,between}} \vee mO_{\text{N,between}} \vee mO_{\text{NE,between}} \vee$$
$$mO_{\text{same,between}} \vee mO_{\text{E,between}} \vee mO_{\text{S,between}} \vee mO_{\text{SE,between}}$$

$$dO_{\text{N,between}} \wedge mO_{\text{SE,down}} = I \times \{\text{before}\} \times I \wedge \{\text{meet}\} \times \{\text{metby}\} \times \{\text{metby}\}$$
$$\rightarrow \{\text{meet}, \text{contain}, \text{finishedby}, \text{overlap}\} \times \{\text{before}, \text{meet}, \text{overlap},$$
$$\text{start}, \text{during}\} \times \{\text{metby}, \text{contain}, \text{startedby}, \text{overlappedby}\}$$
$$\rightarrow dO_{\text{N,between}} \vee dO_{\text{NE,between}} \vee mO_{\text{N,between}} \vee mO_{\text{NE,between}} \vee$$
$$O_{\text{same,between}} \vee mO_{\text{E,between}} \vee mO_{\text{S,between}} \vee mO_{\text{SE,between}} \vee$$
$$dO_{\text{N,down}} \vee dO_{\text{NE,down}} \vee mO_{\text{N,down}} \vee mO_{\text{NE,down}} \vee$$
$$mO_{\text{same,down}} \vee mO_{\text{E,down}} \vee mO_{\text{S,down}} \vee mO_{\text{SE,down}}$$

$$dO_{\text{N,down}} \wedge mO_{\text{SE,up}} = (I \times \{\text{before}, \text{meet}\} \times \{\text{after}\} \bigcup I \times \{\text{before}\} \times \{\text{metby}\}) \wedge$$

$\{\text{meet}\} \times \{\text{metby}\} \times \{\text{meet}\}$

$\rightarrow \{\text{meet}, \text{contain}, \text{finishedby}, \text{overlap}\} \times \{\text{before}, \text{meet}, \text{overlap},$
$\text{start}, \text{during}, \text{finishedby}, \text{equal}, \text{finish}\} \times \{\text{after}, \text{metby},$
$\text{overlappedby}, \text{finish}, \text{during}\} \bigcup \{\text{meet}, \text{contain}, \text{finishedby},$
$\text{overlap}\} \times \{\text{before}, \text{meet}, \text{overlap}, \text{start}, \text{during}\} \times$
$\{\text{startedby}, \text{equal}, \text{start}\}$

$\rightarrow dO_{\text{N,down}} \vee dO_{\text{NE,down}} \vee dO_{\text{same,down}} \vee dO_{\text{E,down}} \vee dO_{\text{S,down}} \vee$
$dO_{\text{SE,down}} \vee dO_{\text{N,between}} \vee dO_{\text{NE,between}} \vee dO_{\text{N,up}} \vee dO_{\text{NE,up}} \vee$
$mO_{\text{N,up}} \vee mO_{\text{NE,up}} \vee mO_{\text{same,up}} \vee mO_{\text{E,up}} \vee mO_{\text{S,up}} \vee mO_{\text{SE,up}} \vee$
$mO_{\text{N,between}} \vee mO_{\text{NE,between}} \vee O_{\text{same,between}} \vee mO_{\text{E,between}} \vee$
$mO_{\text{S,between}} \vee mO_{\text{SE,between}} \vee mO_{\text{N,down}} \vee mO_{\text{NE,down}} \vee$
$mO_{\text{same,down}} \vee mO_{\text{E,down}} \vee mO_{\text{S,down}} \vee mO_{\text{SE,down}}$

$dO_{\text{N,down}} \wedge mO_{\text{SE,between}} = (I \times \{\text{before}, \text{meet}\} \times \{\text{after}\} \bigcup I \times \{\text{before}\} \times \{\text{metby}\}) \wedge$
$\{\text{meet}\} \times \{\text{metby}\} \times I$

$\rightarrow \{\text{meet}, \text{contain}, \text{finishedby}, \text{overlap}\} \times \{\text{before}, \text{meet},$
$\text{overlap}, \text{start}, \text{during}, \text{equal}, \text{finishedby}, \text{finish}\} \times$
$\{\text{after}\} \bigcup \{\text{meet}, \text{contain}, \text{finishedby}, \text{overlap}\} \times$
$\{\text{before}, \text{meet}, \text{overlap}, \text{start}, \text{during}\} \times \{\text{after}, \text{metby}\}$

$\rightarrow dO_{\text{N,down}} \vee dO_{\text{NE,down}} \vee dO_{\text{same,down}} \vee dO_{\text{E,down}} \vee dO_{\text{S,down}} \vee$
$dO_{\text{SE,down}} \vee mO_{\text{N,down}} \vee mO_{\text{NE,down}} \vee mO_{\text{same,down}} \vee$
$mO_{\text{E,down}} \vee mO_{\text{S,down}} \vee mO_{\text{SE,down}}$

表 6.11 $dO_{i_1,j_1} \wedge mO_{i_2,j_2}$

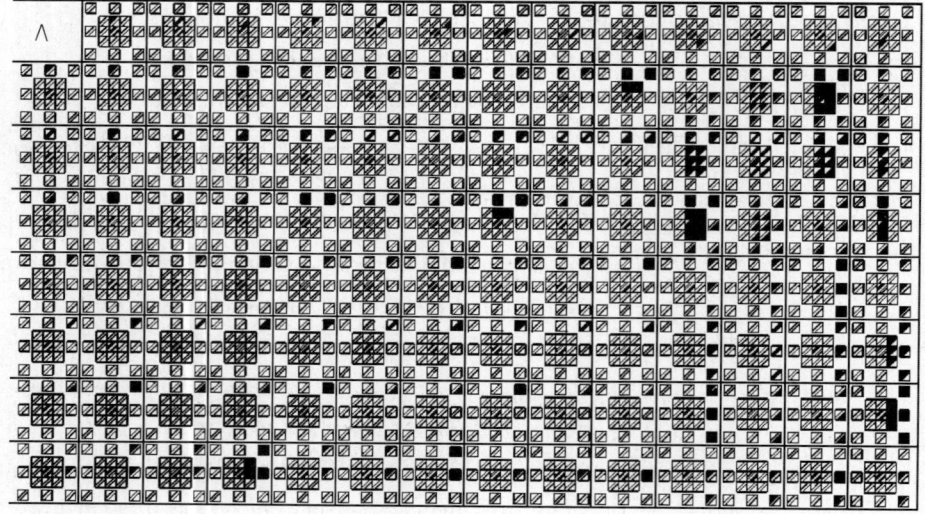

$dO_{N,down} \wedge mO_{SE,down} = (I \times \{before, meet\} \times \{after\} \bigcup I \times \{before\} \times \{metby\}) \wedge \{meet\} \times \{metby\} \times \{metby\}$

$\rightarrow \{meet, contain, finishedby, overlap\} \times \{before, meet, overlap, start, during, finishedby, equal, finish\} \times \{after\}$

$\rightarrow dO_{N,down} \vee dO_{NE,down} \vee dO_{same,down} \vee dO_{E,down} \vee dO_{S,down} \vee dO_{SE,down}$

以上 9 种关系的推理结果用表 6.10 表示,表中 $j_1, j_2 \in DIRZ$。

表 6.10　$dO_{N,j_1} \wedge mO_{SE,j_2} \rightarrow topdir(A, C)$

$dO_{i_1,j_1} \wedge mO_{i_2,j_2} \rightarrow topdir(A, C)$ 的推理结果见表 6.11。表中 $j_1, j_2 \in DIRZ$, $i_1, i_2 \in DIRXY$。

$\rightarrow topdir(A, C)$

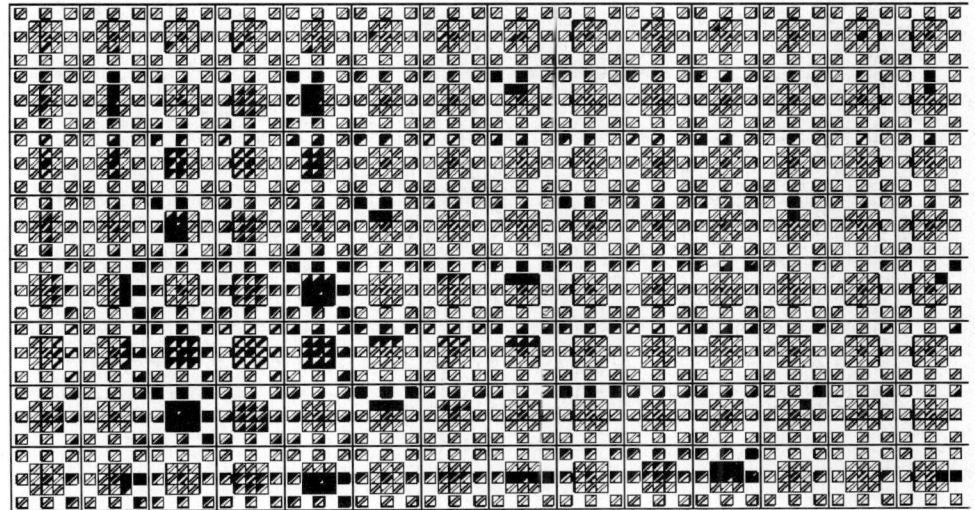

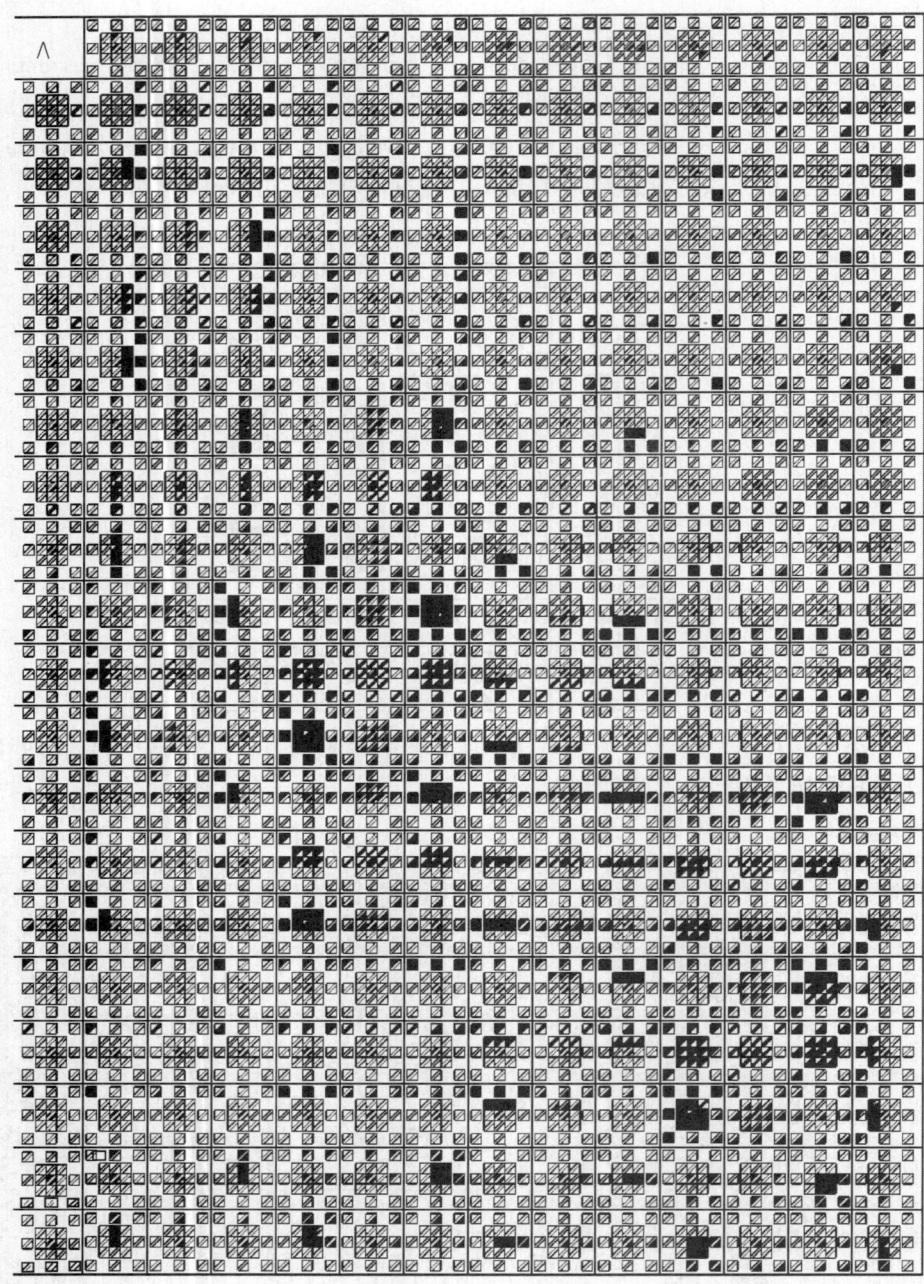

第 6 章 综合拓扑方向关系描述及其推理 135

续表

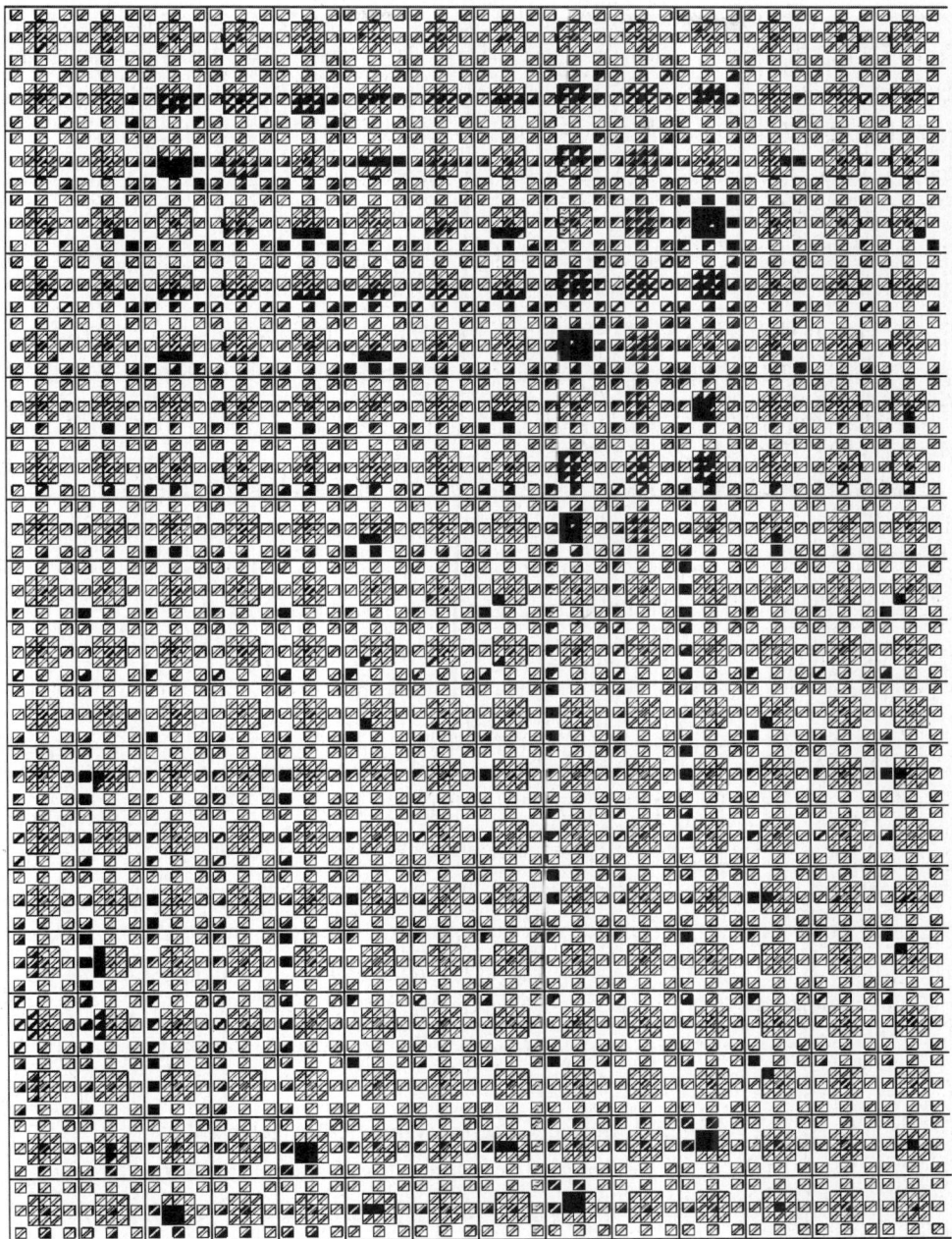

根据表 6.11 可得,对给定的 $i_1,i_2 \in DIRXY, j_1,j_2 \in DIRZ$,由 $dO_{i_1,j_1} \wedge dO_{i_2,j_2} \to topdir(A,C)$ 得出的方向关系 $dir(A,C)$ 与 $O_{i_1,j_1} \wedge O_{i_2,j_2} \to dir(A,C)$ 得出的方向关系 $dir(A,C)$ 相等。当 $i_1 = i_2$ 或 $j_1 = j_2$ 时,A 与 C 的拓扑关系为 disjoint,而且,只要 $dir(A,B)$ 与 $dir(B,C)$ 在由 X,Y 或 Z 轴确定的方向关系上有共同之处,A 与 C 的拓扑关系就为 disjoint,如 $O_{N,up}$ 和 $O_{NE,down}$ 分别属于由 Y 轴确定的 $\{(x,y,z)|y > \max\{y_A\}, (x_A,y_A,z_A) \in A\}$ 和 $\{(x,y,z)|y > \max\{y_B\}, (x_B,y_B,z_B) \in B\}$ 时,A 与 C 的拓扑关系为 disjoint。当 $i_1,i_2 \in \{N,E,S,W\}$ 时,只要 $i_1 \neq i_2$,就有 $dO_{i_1,up} \wedge dO_{i_2,down} = dO_{i_1,down} \wedge dO_{i_2,up}$,且 $top(A,C) =$ disjoint $\vee$ meet;其他情况时,$top(A,C) =$ disjoint。

3. 当拓扑关系 $top(A,B) =$ meet 且 $top(B,C) =$ disjoint 时

当 A 与 B 的拓扑方向关系中的拓扑关系为 meet 时,B 与 C 的拓扑方向关系中的拓扑关系为 disjoint 时,可以推导 A 与 C 的拓扑方向关系,用公式描述为 $mdir(A,B) \wedge ddir(A,B) \to topdir(A,C)$。

$mO_{N,up} \wedge dO_{SE,up} = I \times \{meet\} \times \{meet\} \wedge (\{before, meet\} \times \{after, metby\} \times \{before\} \cup \{before\} \times \{after, metby\} \times \{meet\} \cup \{meet\} \times \{after\} \times \{meet\})$

$\to \{before, meet, contain, finishedby, overlap\} \times \{after, metby, overlappedby, finishedby, contain, startedby, equal, finish\} \times \{before\}$

$\to dO_{same,up} \vee dO_{E,up} \vee dO_{S,up} \vee dO_{SE,up}$

$mO_{N,up} \wedge dO_{SE,between} = I \times \{meet\} \times \{meet\} \wedge (\{before, meet\} \times \{after\} \times I \cup \{before\} \times \{metby\} \times I)$

$\to \{before, meet, contain, finishedby, overlap\} \times \{after, metby, overlappedby, finishedby, contain, startedby, equal, finish\} \times \{before, meet\}$

$\to dO_{same,up} \vee dO_{E,up} \vee dO_{S,up} \vee dO_{SE,up} \vee mO_{same,up} \vee mO_{E,up} \vee mO_{S,up} \vee mO_{SE,up}$

$mO_{N,up} \wedge dO_{SE,down} = I \times \{meet\} \times \{meet\} \wedge (\{before, meet\} \times \{after, metby\} \times \{after\} \cup \{before\} \times \{after, metby\} \times \{after, metby\} \cup \{meet\} \times \{after\} \times \{after\})$

$\to \{before, meet, contain, finishedby, overlap\} \times \{after, metby, overlappedby, finishedby, contain, startedby, equal, finish\} \times \{after, metby, overlappedby, startedby, contain, equal, finishedby, finish\}$

$\to dO_{E,between} \vee dO_{S,between} \vee dO_{SE,between} \vee O_{same,between} \vee mO_{E,between} \vee mO_{S,between} \vee mO_{SE,between} \vee dO_{same,down} \vee dO_{E,down} \vee dO_{S,down} \vee$

$$dO_{\text{SE,down}} \vee mO_{\text{same,down}} \vee mO_{\text{E,down}} \vee mO_{\text{S,down}} \vee mO_{\text{SE,down}}$$

$mO_{\text{N,between}} \wedge dO_{\text{SE,up}} = I \times \{\text{meet}\} \times I \wedge (\{\text{before,meet}\} \times \{\text{after,metby}\} \times \{\text{before}\} \bigcup$
$\{\text{before}\} \times \{\text{after,metby}\} \times \{\text{meet}\} \bigcup \{\text{meet}\} \times \{\text{after}\} \times \{\text{meet}\})$
$\rightarrow \{\text{before,meet,contain,finishedby,overlap}\} \times$
$\{\text{after,metby,overlappedby,finishedby,contain,startedby,}$
$\text{equal,finish}\} \times \{\text{before,meet,contain,finishedby,overlap}\}$
$\rightarrow dO_{\text{same,up}} \vee dO_{\text{E,up}} \vee dO_{\text{S,up}} \vee dO_{\text{SE,up}} \vee mO_{\text{same,up}} \vee mO_{\text{E,up}} \vee$
$mO_{\text{S,up}} \vee mO_{\text{SE,up}} \vee dO_{\text{E,between}} \vee dO_{\text{S,between}} \vee dO_{\text{SE,between}} \vee$
$O_{\text{same,between}} \vee mO_{\text{E,between}} \vee mO_{\text{S,between}} \vee mO_{\text{SE,between}}$

$mO_{\text{N,between}} \wedge dO_{\text{SE,between}} = I \times \{\text{meet}\} \times I \wedge (\{\text{before,meet}\} \times \{\text{after}\} \times I \bigcup \{\text{before}\} \times$
$\{\text{metby}\} \times I)$
$\rightarrow \{\text{before,meet,contain,finishedby,overlap}\} \times$
$\{\text{after,metby,overlappedby,contain,startedby,}$
$\text{equal,startedby,equal,finishedby,finish}\} \times I$
$\rightarrow dO_{\text{E,between}} \vee dO_{\text{S,between}} \vee dO_{\text{SE,between}} \vee O_{\text{same,between}} \vee$
$mO_{\text{E,between}} \vee mO_{\text{S,between}} \vee mO_{\text{SE,between}}$

$mO_{\text{N,between}} \wedge dO_{\text{SE,down}} = I \times \{\text{meet}\} \times I \wedge (\{\text{before,meet}\} \times \{\text{after,metby}\} \times$
$\{\text{after}\} \bigcup \{\text{before}\} \times \{\text{after,metby}\} \times \{\text{after,metby}\} \bigcup$
$\{\text{meet}\} \times \{\text{after}\} \times \{\text{after}\})$
$\rightarrow \{\text{before,meet,contain,finishedby,overlap}\} \times$
$\{\text{after,metby,overlappedby,startedby,contain,equal,}$
$\text{finishedby,finish}\} \times \{\text{after,metby,overlappedby,}$
$\text{startedby,contain}\}$
$\rightarrow dO_{\text{E,between}} \vee dO_{\text{S,between}} \vee dO_{\text{SE,between}} \vee O_{\text{same,between}} \vee mO_{\text{E,between}} \vee$
$mO_{\text{S,between}} \vee mO_{\text{SE,between}} \vee dO_{\text{same,down}} \vee dO_{\text{E,down}} \vee dO_{\text{S,down}} \vee$
$dO_{\text{SE,down}} \vee mO_{\text{same,down}} \vee mO_{\text{E,down}} \vee mO_{\text{S,down}} \vee mO_{\text{SE,down}}$

$mO_{\text{N,down}} \wedge dO_{\text{SE,up}} = I \times \{\text{meet}\} \times \{\text{metby}\} \wedge (\{\text{before,meet}\} \times \{\text{after,metby}\} \times \{\text{before}\} \bigcup$
$\{\text{before}\} \times \{\text{after,metby}\} \times \{\text{meet}\} \bigcup \{\text{meet}\} \times \{\text{after}\} \times \{\text{meet}\})$
$\rightarrow \{\text{before,meet,contain,finishedby,overlap}\} \times$
$\{\text{after,metby,overlappedby,startedby,contain,equal,}$
$\text{finishedby,finish}\} \times \{\text{before,meet,overlap,contain,}$
$\text{finishedby,startedby,equal,start}\}$
$\rightarrow dO_{\text{same,up}} \vee dO_{\text{E,up}} \vee dO_{\text{S,up}} \vee dO_{\text{SE,up}} \vee mO_{\text{same,up}} \vee mO_{\text{E,up}} \vee$
$mO_{\text{S,up}} \vee mO_{\text{SE,up}} \vee dO_{\text{E,between}} \vee dO_{\text{S,between}} \vee dO_{\text{SE,between}} \vee$
$O_{\text{same,between}} \vee mO_{\text{E,between}} \vee mO_{\text{S,between}} \vee mO_{\text{SE,between}}$

$$mO_{\text{N,down}} \wedge dO_{\text{SE,between}} = I \times \{\text{meet}\} \times \{\text{metby}\} \wedge (\{\text{before}, \text{meet}\} \times \{\text{after}\} \times I \cup \{\text{before}\} \times \{\text{metby}\} \times I)$$

$$\rightarrow \{\text{before}, \text{meet}, \text{contain}, \text{finishedby}, \text{overlap}\} \times \{\text{after}, \text{metby}, \text{overlappedby}, \text{startedby}, \text{contain}, \text{equal}, \text{finishedby}, \text{finish}\} \times \{\text{after}, \text{metby}\}$$

$$\rightarrow dO_{\text{same,down}} \vee dO_{\text{E,down}} \vee dO_{\text{S,down}} \vee dO_{\text{SE,down}} \vee mO_{\text{same,down}} \vee mO_{\text{E,down}} \vee mO_{\text{S,down}} \vee mO_{\text{SE,down}}$$

$$mO_{\text{N,down}} \wedge dO_{\text{SE,down}} = I \times \{\text{meet}\} \times \{\text{metby}\} \wedge (\{\text{before}, \text{meet}\} \times \{\text{after}, \text{metby}\} \times \{\text{after}\} \cup \{\text{before}\} \times \{\text{after}, \text{metby}\} \times \{\text{after}\} \cup \{\text{meet}\} \times \{\text{after}\} \times \{\text{after}\})$$

$$\rightarrow \{\text{before}, \text{meet}, \text{contain}, \text{finishedby}, \text{overlap}\} \times \{\text{after}, \text{metby}, \text{overlappedby}, \text{startedby}, \text{contain}, \text{equal}, \text{finishedby}, \text{finish}\} \times \{\text{after}\}$$

$$\rightarrow dO_{\text{same,down}} \vee dO_{\text{E,down}} \vee dO_{\text{S,down}} \vee dO_{\text{SE,down}}$$

表 6.12 给出了上述 9 种关系的推理结果，表中 $j_1, j_2 \in DIRZ$。

表 6.13　$mO_{i_1,j_1} \wedge dO_{i_2,j_2}$

第 6 章　综合拓扑方向关系描述及其推理　　　139

表 6.12　$mO_{N,j_1} \wedge dO_{SE,j_2} \rightarrow topdir(A,C)$

定性推理 $mO_{i_1,j_1} \wedge dO_{i_2,j_2} \rightarrow topdir(A,C)$ 的结果见表 6.13。表中 $i_1, i_2 \in DIRXY$, $j_1, j_2 \in DIRZ$。

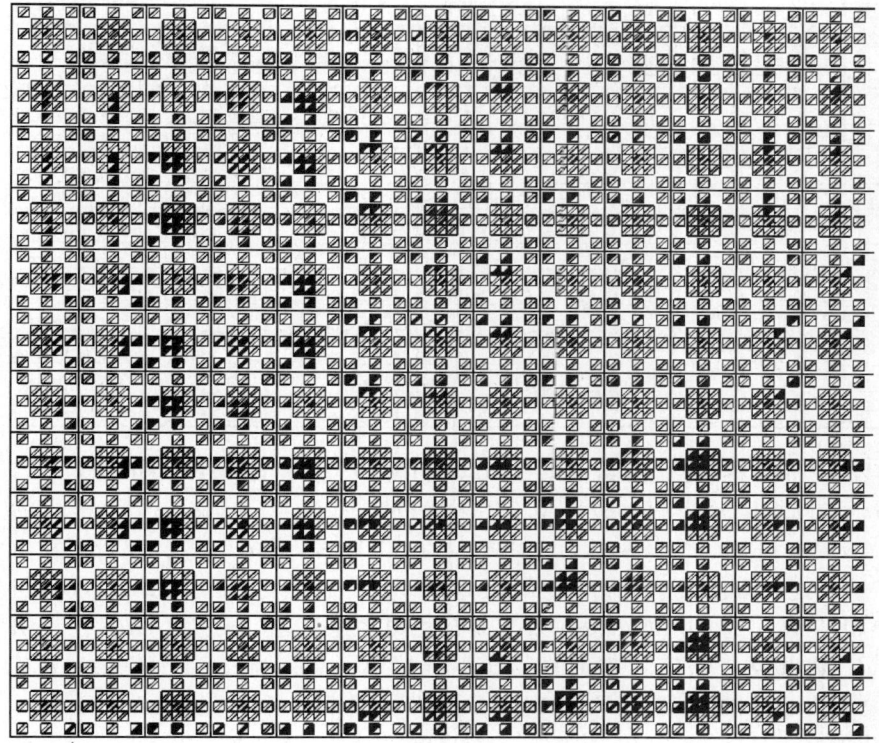

根据表 6.13 可得，对任意给定的 $i_1, i_2 \in DIRXY, j_1, j_2 \in DIRZ$，由 $mO_{i_1,j_1} \wedge dO_{i_2,j_2} \rightarrow topdir(A,C)$ 推出的 $dir(A,C)$ 与由 $O_{i_1,j_1} \wedge O_{i_2,j_2} \rightarrow dir(A,C)$ 推出的 $dir(A,C)$ 不一定相同。当 $i_1 \in \{N, E, S, W\}, i_2 \in \{NE, SE, SW, NW\}$，且两者在由 X 轴和 Y 轴确定的方向关系有相同之处时，由 $mO_{i_1,j_1} \wedge dO_{i_2,j_2}$ 推出的 $dir(A,C)$ 与由 $O_{i_1,j_1} \wedge O_{i_2,j_2}$ 推出的 $dir(A,C)$ 相等；当没有共同之处时，由 $mO_{i_1,j_1} \wedge dO_{i_2,j_2}$ 推出的 $dir(A,C)$ 是由 $O_{i_1,j_1} \wedge O_{i_2,j_2}$ 推出的 $dir(A,C)$ 的真子集。当 $i_1 \in DIRXY$ 和 $i_2 \in DIRXY$ 是互为相反的 N 和 S、E 和 W、NE 和 SW 或 NW 和 SE 时，由 $mO_{i_1,j_1} \wedge dO_{i_2,j_2}$ 推出的 $dir(A,C)$ 是由 $O_{i_1,j_1} \wedge O_{i_2,j_2}$ 推出的 $dir(A,C)$ 的真子集。当 $i_1 = i_2$ 或

续表

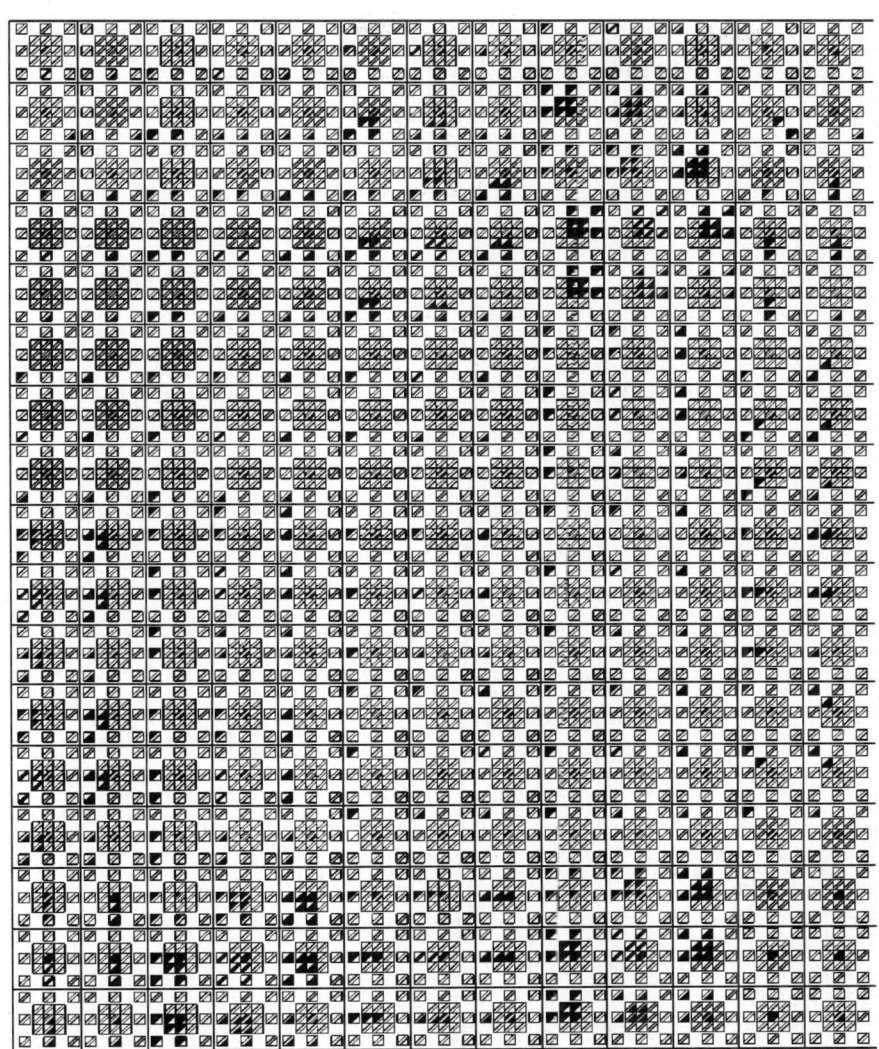

$j_1 = j_2$ 时，由 $mO_{i_1,j_1} \wedge dO_{i_2,j_2} \to topdir(A,C)$ 推出的 $top(A,C) =$ disjoint。

4. 当拓扑关系为 $top(A,B) =$ meet 且 $tcp(B,C) =$ meet 时

当 A 与 B 以及 B 与 C 的拓扑方向关系中的拓扑关系均为 meet 相切时，可以推导 A 与 C 的拓扑方向关系，用公式描述为 $mdir(A,B) \wedge mdir(B,C) \to topdir(A,C)$。下面给出具体推理。

$mO_{N,up} \wedge mO_{E,up} = I \times \{\text{meet}\} \times \{\text{meet}\} \wedge \{\text{meet}\} \times I \times \{\text{meet}\}$
$\to \{\text{meet, contain, finishedby, overlap}\} \times \{\text{before, meet}\} \times \{\text{before}\}$
$\to dO_{N,up} \vee dO_{NE,up}$

$mO_{N,up} \wedge mO_{E,between} = I \times \{meet\} \times \{meet\} \wedge \{meet\} \times I \times I$
$\rightarrow \{meet, contain, finishedby, overlap\} \times \{before, meet\} \times$
$\{before, meet\}$
$\rightarrow dO_{N,up} \vee dO_{NE,up} \vee mO_{N,up} \vee mO_{NE,up}$

$mO_{N,up} \wedge mO_{E,down} = I \times \{meet\} \times \{meet\} \wedge \{meet\} \times I \times \{metby\}$
$\rightarrow \{meet, contain, finishedby, overlap\} \times \{before, meet\} \times$
$\{equal, finishedby, overlappedby\}$
$\rightarrow dO_{N,between} \vee dO_{NE,between} \vee mO_{N,between} \vee mO_{NE,between} \vee$
$dO_{N,down} \vee dO_{NE,down} \vee mO_{N,down} \vee mO_{NE,between}$

$mO_{N,between} \wedge mO_{E,up} = I \times \{meet\} \times I \wedge \{meet\} \times I \times \{meet\}$
$\rightarrow \{meet, contain, finishedby, overlap\} \times \{before, meet\} \times$
$\{meet, contain, finishedby, overlap\}$
$\rightarrow dO_{N,up} \vee dO_{N,between} \vee dO_{NE,up} \vee dO_{NE,between} \vee mO_{N,up} \vee$
$mO_{N,between} \vee mO_{NE,up} \vee mO_{NE,between}$

$mO_{N,between} \wedge mO_{E,between} = I \times \{meet\} \times I \wedge \{meet\} \times I \times I$
$\rightarrow \{meet, contain, finishedby, overlap\} \times \{before, meet\} \times I$
$\rightarrow dO_{N,between} \vee dO_{NE,between} \vee mO_{N,between} \vee mO_{NE,between}$

$mO_{N,between} \wedge mO_{E,down} = I \times \{meet\} \times I \wedge \{meet\} \times I \times \{metby\}$
$\rightarrow \{meet, contain, finishedby, overlap\} \times \{before, meet\} \times$
$\{contain, startedby, overlappedby, metby\}$
$\rightarrow dO_{N,between} \vee dO_{N,down} \vee dO_{NE,between} \vee dO_{NE,down} \vee mO_{N,between} \vee$
$mO_{N,down} \vee mO_{NE,between} \vee mO_{NE,down}$

$mO_{N,down} \wedge mO_{E,up} = I \times \{meet\} \times \{metby\} \wedge \{meet\} \times I \times \{meet\}$
$\rightarrow \{meet, contain, finishedby, overlap\} \times \{before, meet\} \times$
$\{startedy, equal, start\}$

表 6.15　$mO_{i_1,j_1} \wedge mO_{i_2,j_2}$

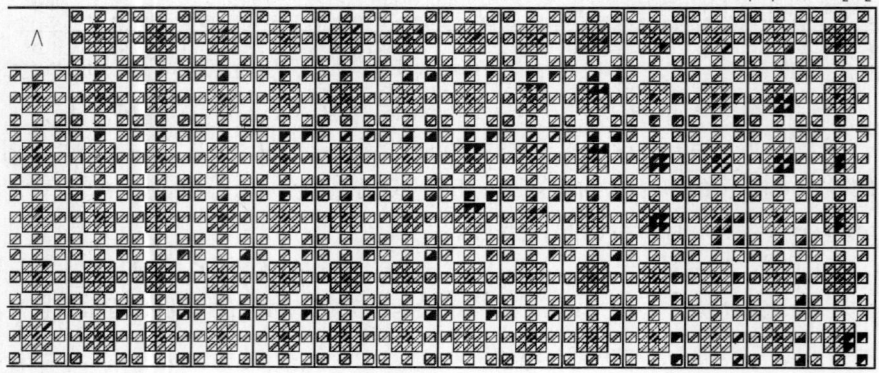

$\rightarrow dO_{N,between} \lor dO_{N,up} \lor dO_{NE,between} \lor dO_{NE,up} \lor mO_{N,between} \lor mO_{N,up} \lor mO_{NE,between} \lor mO_{NE,up}$

$mO_{N,down} \land mO_{E,between} = I \times \{meet\} \times \{meetby\} \land \{meet\} \times I \times I$
$\rightarrow \{meet, contain, finishedby, overlap\} \times \{before, meet\} \times \{after, metby\}$
$\rightarrow dO_{N,down} \lor dO_{NE,down} \lor mO_{N,down} \lor mO_{NE,down}$

$mO_{N,down} \land mO_{E,down} = I \times \{meet\} \times \{metby\} \land \{meet\} \times I \times \{metby\}$
$\rightarrow \{meet, contain, finishedby, overlap\} \times \{before, meet\} \times \{after\}$
$\rightarrow dO_{N,down} \lor dO_{NE,down}$

用表 6.14 表示上述 9 个关系的定性推理结果,表中 $j_1, j_2 \in DIRZ$。

表 6.14　$mO_{N,j_1} \land mO_{E,j_2} \rightarrow topdir(A,C)$

$mO_{i_1,j_1} \land mO_{i_2,j_2} \rightarrow topdir(A,C)$ 的推理结果见表 6.15。表中 $i_1, i_2 \in DIRXY$,$j_1, j_2 \in DIRZ$。

$\rightarrow topdir(A,C)$

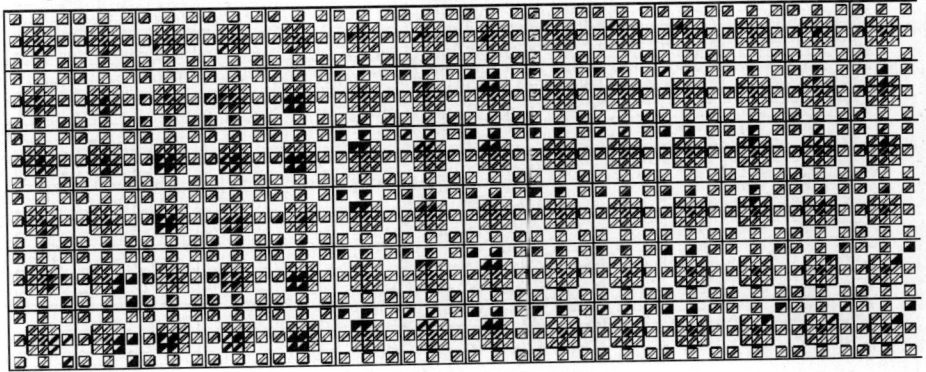

续表

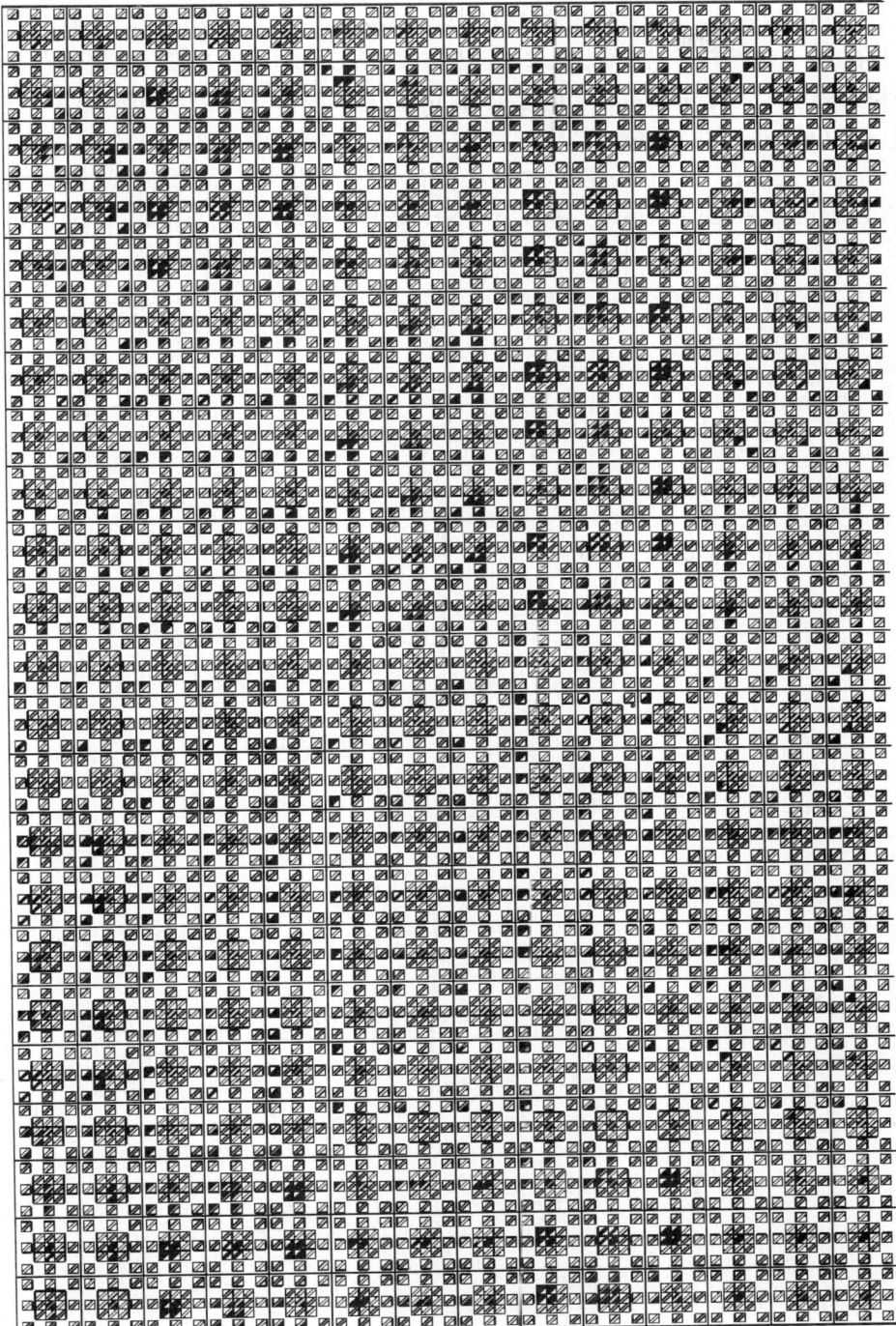

根据表 6.15 可得,对任意给定的 $i_1,i_2 \in DIRXY, j_1,j_2 \in DIRZ$,由 $mO_{i_1,j_1} \wedge dO_{i_2,j_2} \to topdir(A,C)$ 推出的 $dir(A,C)$ 与由 $O_{i_1,j_1} \wedge O_{i_2,j_2} \to dir(A,C)$ 推出的 $dir(A,C)$ 不一定相同。当 $i_1 \in \{N,E,S,W\}, i_2 \in \{NE,SE,SW,NW\}$,且两者在由 X 轴和 Y 轴确定的方向关系有相同之处时,由 $mO_{i_1,j_1} \wedge dO_{i_2,j_2}$ 推出的 $dir(A,C)$ 与由 $O_{i_1,j_1} \wedge O_{i_2,j_2}$ 推出的 $dir(A,C)$ 相同,当没有共同之处时,由 $mO_{i_1,j_1} \wedge dO_{i_2,j_2}$ 推出的 $dir(A,C)$ 是由 $O_{i_1,j_1} \wedge O_{i_2,j_2}$ 推出的 $dir(A,C)$ 的真子集。当 $i_1 \in DIRXY$ 和 $i_2 \in DIRXY$ 是互为相反的 N 和 S、E 和 W、NE 和 SW 或 NW 和 SE 时,$mO_{i_1,j_1} \wedge dO_{i_2,j_2}$ 推出的 $dir(A,C)$ 是由 $O_{i_1,j_1} \wedge O_{i_2,j_2}$ 推出的 $dir(A,C)$ 的真子集。当 $i_1 = i_2$ 或 $j_1 = j_2$ 时,由 $mO_{i_1,j_1} \wedge dO_{i_2,j_2} \to topdir(A,C)$ 推出的 $top(A,C) = \text{disjoint}$。

第7章 位置信息的定性描述及其定性推理

目标对象与参照物之间的位置关系可用综合拓扑方向关系描述。综合拓扑方向关系描述方法只在小尺度环境下对研究对象的位置信息提供帮助(Hernández，1994；Sharma et al,1995)，如"房间中的物体"可用拓扑关系"contain"和方向关系"same"表示。在大尺度环境下，当方向关系相同，且拓扑关系均为"disjoint"时，两者的拓扑方向关系相同，如图 7.1 所示。在图 7.1(a)中，B 与 A 的拓扑关系为"disjoint"，方向关系为"NE"；在图 7.1(b)中，B 与 A 的拓扑关系为"disjoint"，方向关系为"NE"。用拓扑关系和方向关系表达图 7.1 中 B 和 A 的位置关系都是"disjointNE"，因此，用拓扑方向关系无法区分图 7.1 所示的两种情况。

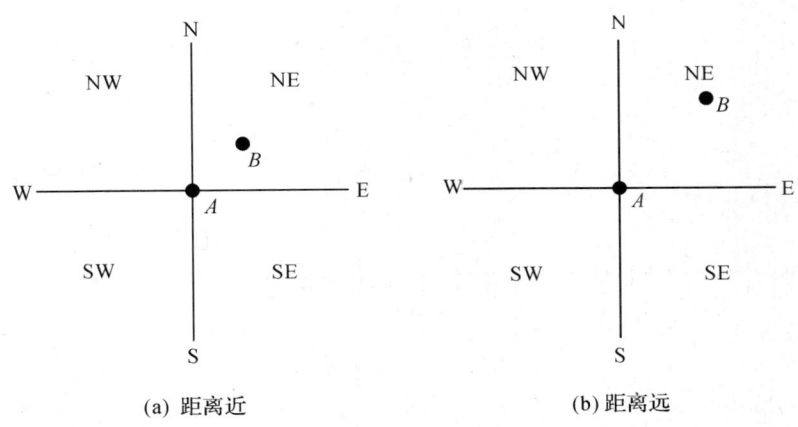

图 7.1 目标对象相对于参照物之间的位置关系

在用定性距离描述目标对象 B 与参照物 A 间的空间关系时，图 7.2(a)和图 7.2(b)所示 A 和 B 之间的定性距离均为 q_2。事实上，图 7.2(a)和图 7.2(b)中目标对象和参照物之间的空间关系差别很大，单纯用定性距离无法区分两图所示的情况。当用方向关系和定性距离描述目标对象 B 与参照物 A 间的位置关系时，就可以区分图 7.2(a)和图 7.2(b)所示的两种情况，如图 7.3 所示。在图 7.3(a)中，B 与参照物 A 的位置关系为 B 在 A 的东北方向，且与 A 的定性距离为 q_2；在图 7.3(b)中，B 与参照物 A 的位置关系为 B 在 A 的西南方向，且与 A 的定性距离为 q_2。

用方向关系和定性距离描述目标对象相对于参照物位置关系的方法，为位置关系的定性描述。目前对位置关系的定性描述的研究主要集中于二维空间位置关系的定性描述,所采用的方向关系均为基于锥形的方向关系(Hong,1994;Clementini et al,1997b)。本章的研究重点是采用基于投影的方向关系建立位置关系定性描述。分别在二维空间和三维空间中研究位置关系的描述公式,根据该公式研究位置关系的定性推理。

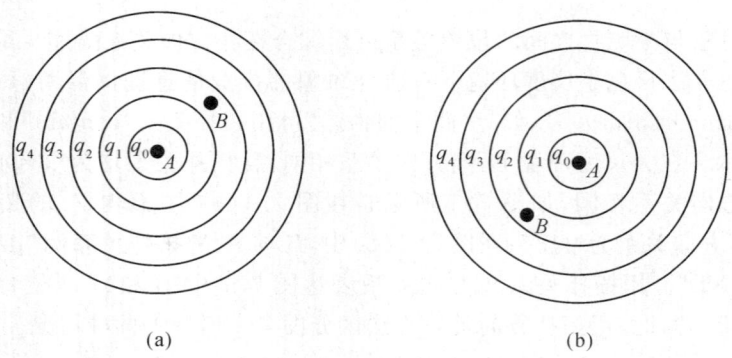

图7.2 被5个定性距离系统划分的二维空间

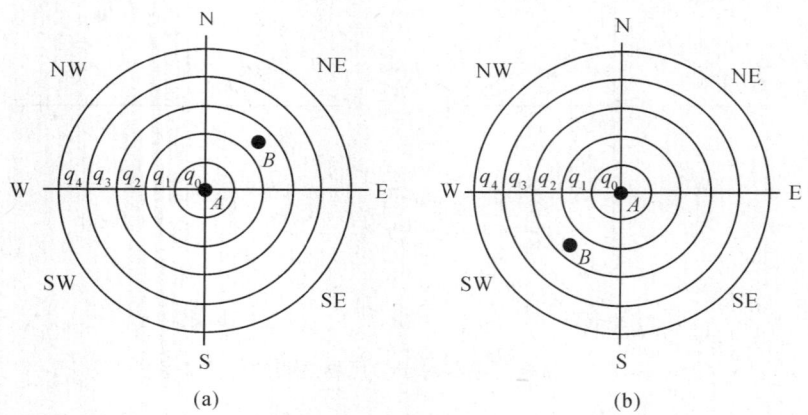

图7.3 由5个定性距离系统和8个方向关系系统划分的二维空间

§7.1 二维位置关系的定性描述及其定性推理

7.1.1 位置关系定性描述

在用定性距离和方向关系综合描述物体的位置时,采用定性距离在前、方向关系在后的方法,记为 $d_{AB}dir(A,B)$,其中,A 为参照对象,B 为目标对象,d_{AB} 为 B 与 A 之间的定性距离,$dir(A,B)$ 为 B 相对于 A 的方向关系。图7.3(a)和图7.3(b)

中 B 相对于 A 的位置信息分别为 $q_2\text{NE}$，$q_2\text{SW}$。本书定性距离描述采用"很近"(q_0)，"近"(q_1)，"中间"(q_2)，"远"(q_3)，"很远"(q_4) 5 个定性距离描述目标对象与参照物间的远近关系，这 5 个定性距离所对应的距离范围分别为：$(0,a_1]$，$(a_1,a_2]$，$(a_2,a_3]$，$(a_3,a_4]$，$(a_4,a_5]$，并且相邻定性距离的距离范围之比为常数 k。方向关系采用投影法描述，并且参照物为点，用八方向关系描述二维空间中目标对象与参照物间的方向关系，如图 7.4 所示。

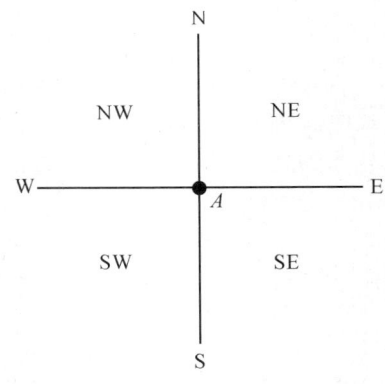

图 7.4 被八方向关系系统划分的二维空间

7.1.2 位置关系定性推理

二维空间中位置关系的定性推理是根据物体 A 与 B 间的定性距离与方向关系，以及 B 与 C 间的定性距离与方向关系，推导 A 与 C 间的位置关系，用公式描述为

$$q_i dir(A,B) \wedge q_j dir(B,C) \rightarrow d_{AC} dir(A,C) \tag{7.1}$$

由于方向关系的定性推理具有传递性，当 $dir(A,B)=dir(B,C)$ 时，C 和 A 的方向关系与 B 和 A 的方向关系相同，即 $dir(A,C)=dir(A,B)$。当 A、B 和 C 在同一条直线上，但 C 和 B 的方向关系与 B 和 A 的方向关系相反时，记为 $dir(B,C)=-dir(A,B)$，式(7.1)写为

$$(A_i,A_{i+1}] - (A_j,A_{j+1}] = (A_i - A_{j+1}, A_{i+1} - A_j) \tag{7.2}$$

(1) 当 $i=j$ 时，$(a_i,a_{i+1}] - (a_i,a_{i+1}] = (a_i - a_{i+1}, a_{i+1} - a_i) = (-\delta_i, \delta_i)$。又因 $\delta_i \in (a_i, a_i + \delta_i] = (a_i, a_{i+1}]$，所以 $(a_i, a_{i+1}] - (a_i, a_{i+1}]$ 所对应的位置关系为：$-q_i$，$-q_{i-1}$，$\cdots$，$-q_0$，q_0，q_1，$\cdots$，q_i，其中，$-q_k$ 表示 $dir(A,C)$ 与 $dir(A,B)$ 相反，且 C 与参照物 A 的定性距离为 q_k，$k=1,2,\cdots,i$。

(2) 当 $i=j+1$ 时，$(a_{j+1},a_{j+2}] - (a_j,a_{j+1}] = (0, a_{j+2} - a_j) = (0, \Delta_{j+1} - a_j) = (0, \delta_{j+1})$。由于 $\delta_{j+1} \in (a_{j+1}, a_{j+1} + \delta_{j+1}] = (a_{j+1}, a_{j+2}]$，所以 $(a_{j+1}, a_{j+2}] - (a_j, a_{j+1}]$ 所对应的位置关系为 $q_0, q_1, \cdots, q_{j+1}$。

(3) 当 $i=j-1$ 时，$(a_i, a_{i+1}] - (a_{i+1}, a_{i+2}] = (a_i - a_{i+2}, 0) = (-\delta_{i+1}, 0)$。所以，$(a_i, a_{i+1}] - (a_{i+1}, a_{i+2}]$ 所对应的位置关系为 $-q_0, -q_1, \cdots, -q_{i+1}$。

上述方法只研究了 A、B 和 C 在一条直线上，且 $dir(B,C) = -dir(A,B)$ 时，A、B 和 C 位置关系的定性推理方法。下面研究 $dir(A,B)$ 和 $dir(B,C)$ 为任意方向关系时的定性推理。

当 B 与 A 的距离为 $d(A,B) \leq \Delta_{i-1}$ 时，B 的范围为

$$\left\{(x_B, y_B) \mid \begin{cases} x_B - x_A = d(A,B)\cos\theta_A & d(A,B) \leqslant \Delta_{i-1} \\ y_B - y_A = d(A,B)\sin\theta_A & 0 \leqslant \theta_A < 2\pi \end{cases}\right\} \quad (7.3)$$

式中,x 和 y 是坐标系 xoy 下的坐标,x 轴指向东为正,y 轴指向北为正;θ_A 为 AB 与 x 轴之间的夹角,逆时针为正,如图 7.5 所示。

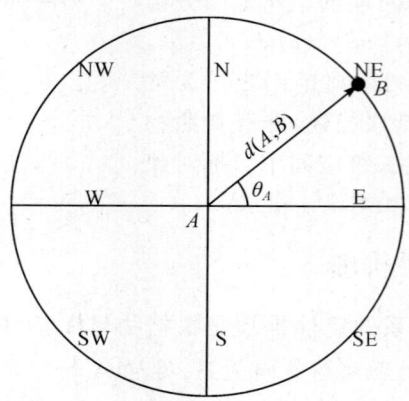

图 7.5 θ_A 的定义

当 B 与 A 的距离为 $d(A,B) \leqslant \Delta_{i-1} + \delta_i$ 时,B 的范围为

$$\left\{(x_B, y_B) \mid \begin{cases} x_B - x_A = d(A,B)\cos\theta_A & d(A,B) \leqslant \Delta_{i-1} + \delta_i \\ y_B - y_A = d(A,B)\sin\theta_A & 0 \leqslant \theta_A < 2\pi \end{cases}\right\} \quad (7.4)$$

当 B 与 A 的距离为 $\Delta_{i-1} < d(A,B) \leqslant \Delta_{i-1} + \delta_i$ 时,B 的范围为

$$\left\{(x_B, y_B) \mid \begin{cases} x_B - x_A = d(A,B)\cos\theta_A & d(A,B) \leqslant \Delta_{i-1} + \delta_i \\ y_B - y_A = d(A,B)\sin\theta_A & 0 \leqslant \theta_A < 2\pi \end{cases}\right\} -$$

$$\left\{(x_B, y_B) \mid \begin{cases} x_B - x_A = d(A,B)\cos\theta_A & d(A,B) \leqslant \Delta_{i-1} \\ y_B - y_A = d(A,B)\sin\theta_A & 0 \leqslant \theta_A < 2\pi \end{cases}\right\}$$

$$= \left\{(x_B, y_B) \mid \begin{cases} x_B - x_A = d(A,B)\cos\theta_A & \Delta_{i-1} < d(A,B) \leqslant \Delta_{i-1} + \delta_i \\ y_B - y_A = d(A,B)\sin\theta_A & 0 \leqslant \theta_A < 2\pi \end{cases}\right\} \quad (7.5)$$

当 C 与 B 的距离为 $\Delta_{j-1} < d(B,C) \leqslant \Delta_{j-1} + \delta_j$ 时,C 的范围为

$$\left\{(x_C, y_C) \mid \begin{cases} x_C - x_B = d(B,C)\cos\theta_B & \Delta_{j-1} < d(B,C) \leqslant \Delta_{j-1} + \delta_j \\ y_C - y_B = d(B,C)\sin\theta_B & 0 \leqslant \theta_B < 2\pi \end{cases}\right\} \quad (7.6)$$

当 $\Delta_{i-1} < d(A,B) \leqslant \Delta_{i-1} + \delta_i$,且 $\Delta_{j-1} < d(B,C) \leqslant \Delta_{j-1} + \delta_j$ 时,目标对象 C 的范围为

$$\left\{(x_B, y_B) \mid \begin{cases} x_B - x_A = d(A,B)\cos\theta_A & \Delta_{i-1} \leqslant d(A,B) \leqslant \Delta_{i-1} + \delta_i \\ y_B - y_A = d(A,B)\sin\theta_A & 0 \leqslant \theta_A < 2\pi \end{cases}\right\} \cap$$

$$\left\{(x_C,y_C) \Bigg| \begin{matrix} x_C-x_B=d(B,C)\cos\theta_B & \Delta_{i-1}\leqslant d(E,C)\leqslant\Delta_{i-1}+\delta_i \\ y_C-y_B=d(B,C)\sin\theta_B & 0\leqslant\theta_B<2\pi \end{matrix}\right\}$$

$$=\left\{(x_C,y_C) \Bigg| \begin{matrix} x_C-x_A=d(A,B)\cos\theta_A+d(B,C)\cos\theta_B & d(A,B)\in(\Delta_{i-1},\Delta_{i-1}+\delta_i] \\ & d(B,C)\in(\Delta_{j-1},\Delta_{j-1}+\delta_j] \\ y_C-y_A=d(A,B)\sin\theta_A+d(B,C)\sin\theta_B & 0\leqslant\theta_A,\theta_B<2\pi \end{matrix}\right\}$$

(7.7)

令 $x_{AC}=x_C-x_A,y_{AC}=y_C-y_A$，根据式(7.7)，$C$ 相对于参照物 A 的坐标范围为

$$\left\{(x_{AC},y_{AC}) \Bigg| \begin{matrix} x_{AC}=d(A,B)\cos\theta_A+d(B,C)\cos\theta_B & d(A,B)\in(\Delta_{i-1},\Delta_{i-1}+\delta_i] \\ & d(B,C)\in(\Delta_{j-1},\Delta_{j-1}+\delta_j] \\ y_{AC}=d(A,B)\sin\theta_A+d(B,C)\sin\theta_B & 0\leqslant\theta_A,\theta_B<2\pi \end{matrix}\right\}$$

(7.8)

当用角度 θ_A 描述目标对象 C 与参照物 A 的方向关系时，8 个方向关系 N、NE、E、SE、S、SW、W、NW 所对应的 θ_A 的范围依次为：$\frac{\pi}{2}$、$(\frac{\pi}{2},0)$、0、$(\frac{3\pi}{2},2\pi)$、$\frac{3\pi}{2}$、$(\pi,\frac{3\pi}{2})$、π、$(\frac{\pi}{2},\pi)$。因此

(1) 当 $\theta_A=\theta_B$ 时，$\left\{(x_{AC},y_{AC}) \Big| \begin{matrix} x_{AC}=(q_i+q_j)\cos\theta_A \\ y_{AC}=(q_i+q_j)\sin\theta_A \end{matrix}\right\}$。

(2) 当 $\theta_B=\theta_A+\pi$ 时，$\left\{(x_{AC},y_{AC}) \Big| \begin{matrix} x_{AC}=(q_i-q_j)\cos\theta_A \\ y_{AC}=(q_i-q_j)\sin\theta_A \end{matrix}\right\}$。

根据式(7.8)可以模拟目标对象 C 相对于参照物 A 的位置关系，图 7.6 是 B 与 A 的位置关系为 $q_1\text{NE}$, $dir(B,C)=\text{SE}$ 时，目标对象 C 相对于 A 的位置关系。

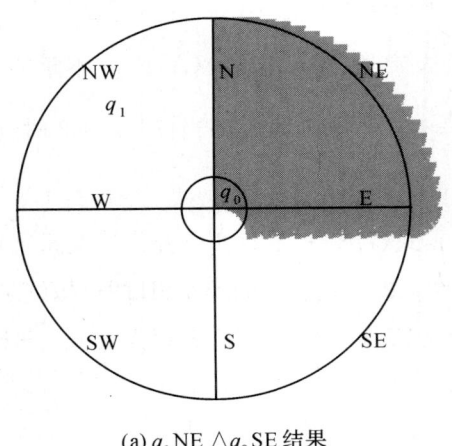

(a) $q_1\text{NE} \wedge q_0\text{SE}$ 结果

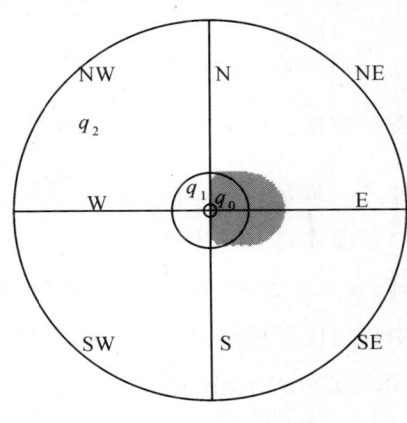

(b) $q_1\text{NE} \wedge q_1\text{SE}$ 结果

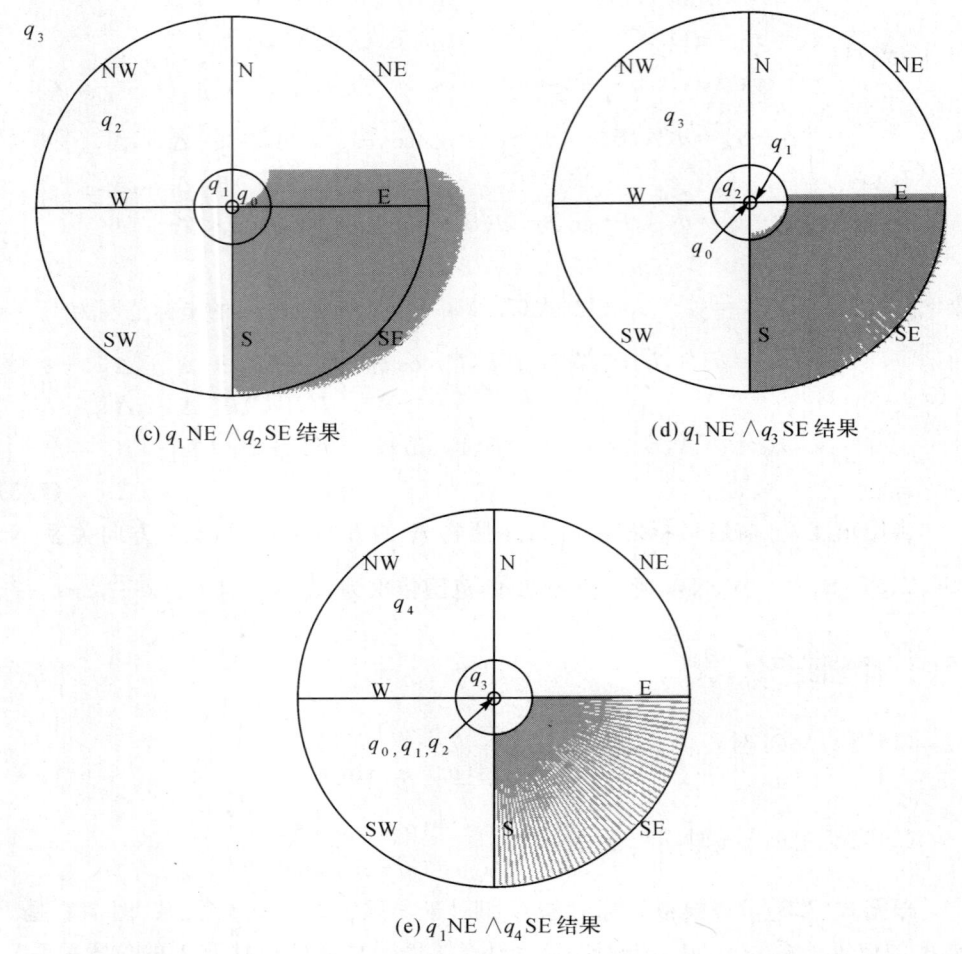

(c) $q_1\text{NE} \wedge q_2\text{SE}$ 结果

(d) $q_1\text{NE} \wedge q_3\text{SE}$ 结果

(e) $q_1\text{NE} \wedge q_4\text{SE}$ 结果

图 7.6　$q_1\text{NE} \wedge q_j\text{SE}$ 的模拟结果

(3) 当 $q_i \gg q_j$ 时,根据吸收律,式(7.1)的推理结果为 $q_i dir(A,B)$。本章的以下部分解释二元关系 $\gg$ 为:取 $\varepsilon = \dfrac{1}{k^2}$,只要 $\dfrac{q_i}{q_j} \leqslant \varepsilon$,就有 $q_i \gg q_j$,即当 $|i-j| \geqslant 2$ 时,就有 $q_i \gg q_j$ 或 $q_i \ll q_j$。

下面以 $dir(A,B) = \text{NE}, k=5$ 为例,根据式(7.8)和吸收律研究位置关系定性推理,表 7.1 为 $q_i\text{NE} \wedge q_j\text{SE}$ 的定性推理结果,图 7.6 为 $q_1\text{NE} \wedge q_j\text{SE}$ 的模拟结果,表中将定性推理结果 $q_0\text{NE} \vee q_0\text{E} \vee q_0\text{SE} \vee q_1\text{NE} \vee q_1\text{E} \vee q_1\text{SE}$ 简写为 $q_0/q_1\{\text{NE},\text{E},\text{SE}\}$,$q_0\text{SE} \vee q_1\text{SE}$ 简写为 $q_1/q_1\text{SE}$,余同。

第 7 章 位置信息的定性描述及其定性推理

表 7.1 q_iNE $\wedge$ q_jSE 推理组合表

$\wedge$	q_0SE	q_1SE	q_2SE	q_3SE	q_4SE
q_0NE	q_0/q_1 {NE, E, SE}	$q_0/q_1/q_2$ {NE,E,SE}	q_2SE	q_3SE	q_4SE
q_1NE	$q_0/q_1/q_2$ {NE,E,SE}	$q_0/q_1/q_2$ {NE,E,SE}	$q_0/q_1/q_2/q_3$ {NE,E,SE}	q_3SE	q_4SE
q_2NE	q_2NE	q_0NE,$q_1/q_2/q_3$ {NE,E,SE}	$q_0/q_1/q_2/q_3$ {NE,E,SE}	q_0SE,$q_1/q_2/q_3/q_4$ {NE,E,SE}	q_4SE
q_3NE	q_3NE	q_3NE	q_0/q_1NE, $q_2/q_3/q_4$ {NE,E,SE}	$q_0/q_1/q_2/q_3/q_4$ {NE,E,SE}	q_0/q_1SE, $q_2/q_3/q_4$ {NE,E,SE}
q_4NE	q_4NE	q_4NE	q_4NE	$q_0/q_1/q_2$NE, q_3/q_4 {NE,E,SE}	$q_0/q_1/q_2/q_3/q_4$ {NE,E,SE}

当 $i=1$,$|i-j|\leqslant 1$ 时,q_1NE $\wedge$ q_jSW 的模拟结果见图 7.7。q_iNE $\wedge$ q_jSW 的推理结果见表 7.2。

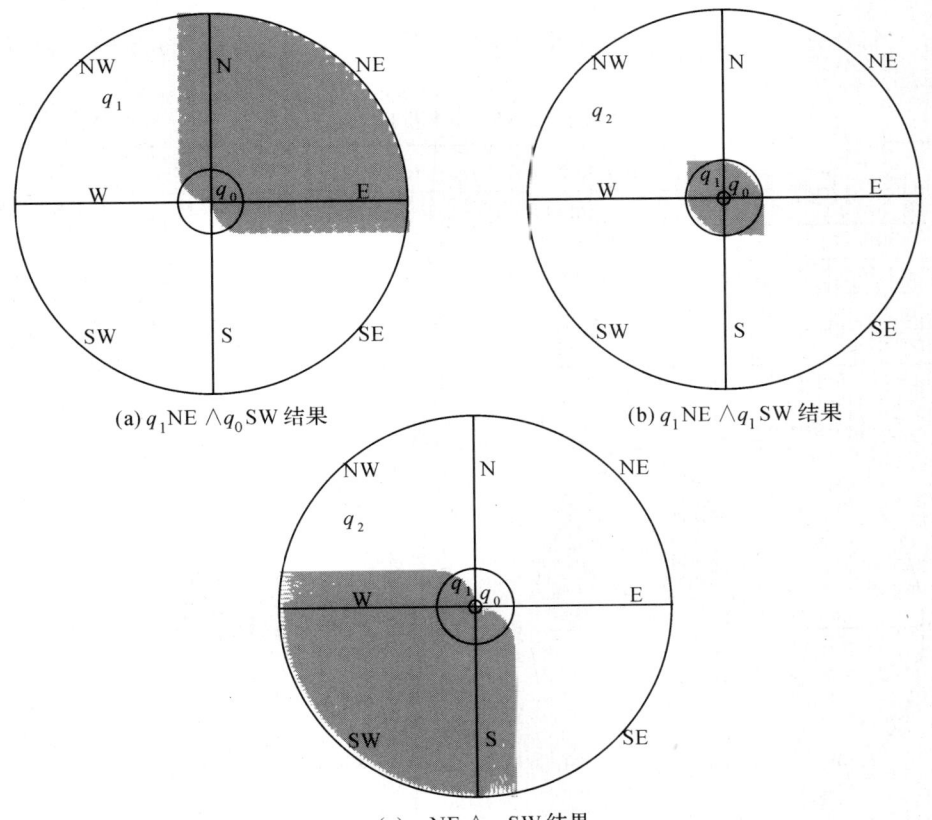

(a) q_1NE $\wedge q_0$SW 结果
(b) q_1NE $\wedge q_1$SW 结果
(c) q_1NE $\wedge q_2$SW 结果

图 7.7 q_1NE $\wedge q_j$SW 的模拟结果($|1-j|\leqslant 1$)

表 7.2 $q_i\text{NE} \wedge q_j\text{SW}$ 推理组合表

$\wedge$	$q_0\text{SW}$	$q_1\text{SW}$	$q_2\text{SW}$	$q_3\text{SW}$	$q_4\text{SW}$
$q_0\text{NE}$	$q_0 D,$ $q_1\{\text{NW,SE}\}$	$q_0/q_1\{\text{SE,S,} \text{SW,W,NW}\}$	$q_2\text{SW}$	$q_3\text{SW}$	$q_4\text{SW}$
$q_1\text{NE}$	$q_0/q_1\{\text{NW,} \text{N,NE,} \text{E,SE}\}$	$q_0/q_1 D,$ $q_2\{\text{NW,SE}\}$	$q_0/q_1/q_2\{\text{SE,} \text{S,SW,W,} \text{NW}\}$	$q_3\text{SW}$	$q_4\text{SW}$
$q_2\text{NE}$	$q_0\text{SW}$	$q_0/q_1/q_2$ $\{\text{NW,N,} \text{NE,E,SE}\}$	$q_0/q_1/q_2 D,$ $q_3\{\text{NW,SE}\}$	$q_0/q_1/q_2/q_3\{\text{SE,} \text{S,SW,W,NW}\}$	$q_4\text{SW}$
$q_3\text{NE}$	$q_0\text{SW}$	$q_1\text{SW}$	$q_0/q_1/q_2/q_3$ $\{\text{NW,N,NE,} \text{E,SE}\}$	$q_0/q_1/q_2/q_3 D,$ $q_4\{\text{NW,SE}\}$	$q_0/q_1/q_2/q_3/q_4$ $\{\text{SE,S,SW,} \text{W,NW}\}$
$q_4\text{NE}$	$q_0\text{SW}$	$q_1\text{SW}$	$q_2\text{SW}$	$q_0/q_1/q_2/q_3/q_4$ $\{\text{NW,N,NE,E,SE}\}$	$q_0/q_1/q_2/$ $q_3/q_4 D$

注:$D=\{\text{N,NE,E,SE,S,SW,W,NW,same}\}$。

当 $|1-j| \leqslant 1$ 时,$q_1\text{NE} \wedge q_j\text{N}$ 的模拟结果见图 7.8。$q_i\text{NE} \wedge q_j\text{N}$ 的推理结果见表 7.3。

表 7.3 $q_i\text{NE} \wedge q_j\text{N}$ 推理组合表

$\wedge$	$q_0\text{N}$	$q_1\text{N}$	$q_2\text{N}$	$q_3\text{N}$	$q_4\text{N}$
$q_0\text{NE}$	$q_0/q_1\text{NE}$	$q_1/q_2\text{NE}$	$q_2\text{N}$	$q_3\text{N}$	$q_4\text{N}$
$q_1\text{NE}$	$q_1/q_2\text{NE}$	$q_1/q_2\text{NE}$	$q_2/q_3\text{NE}$	$q_3\text{N}$	$q_4\text{N}$
$q_2\text{NE}$	$q_2\text{NE}$	$q_2/q_3\text{NE}$	$q_2/q_3\text{NE}$	$q_3/q_4\text{NE}$	$q_4\text{N}$
$q_3\text{NE}$	$q_3\text{NE}$	$q_3\text{NE}$	$q_3/q_4\text{NE}$	$q_3/q_4\text{NE}$	$q_4\text{NE}$
$q_4\text{NE}$	$q_4\text{NE}$	$q_4\text{NE}$	$q_4\text{NE}$	$q_4\text{NE}$	$q_4\text{NE}$

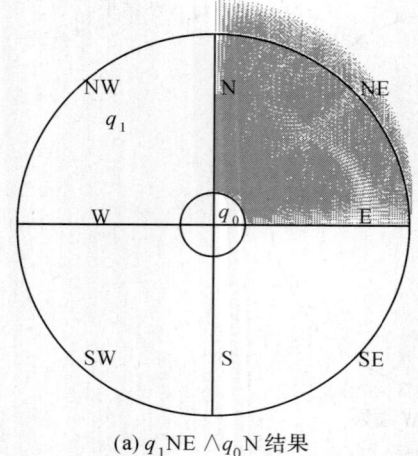

(a) $q_1\text{NE} \wedge q_0\text{N}$ 结果

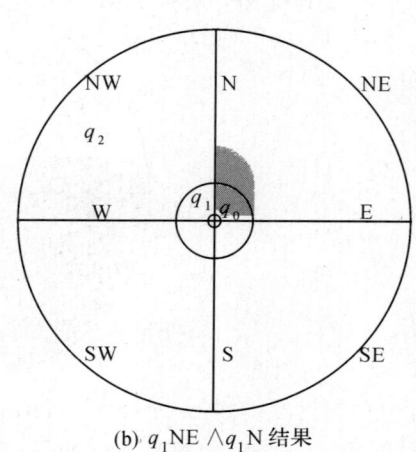

(b) $q_1\text{NE} \wedge q_1\text{N}$ 结果

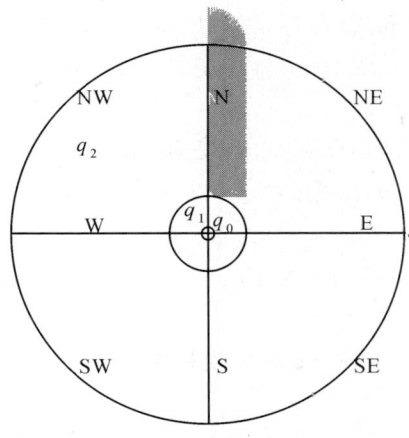

(c) $q_1\mathrm{NE} \wedge q_2\mathrm{N}$ 结果

图 7.8 $q_1\mathrm{NE} \wedge q_j\mathrm{N}$ 的模拟结果($|1-j| \leqslant 1$)

由表 7.3 可以看出,当 $|i-j| \leqslant 1$ 时,$q_i\mathrm{NE} \wedge q_j\mathrm{N} \to q_{\max\{i,j\}}\mathrm{NE} \vee q_{\max\{i,j\}+1}\mathrm{NE}$,定性推理 $q_i\mathrm{NE} \wedge q_{i-1}\mathrm{N}$,$q_i\mathrm{NE} \wedge q_i\mathrm{N}$ 和 $q_{i-1}\mathrm{NE} \wedge q_i\mathrm{N}$ 的结果相同,均为 $q_{\max\{i,j\}}\mathrm{NE} \vee q_{\max\{i,j\}+1}\mathrm{NE}$,即目标对象 C 相对于参照物 A 的位置关系是相同的。但实际上,C 相对于 A 的坐标是不同的,如图 7.8(a)、(b)和图 7.9 所示。根据方向区域的定义可知,方向区域 NE 和 N 在坐标轴 Y 上的方向是相同的,$q_i\mathrm{NE}$ 在 Y 轴上的坐标范围为 $(0, a_{i+1}]$,$q_j\mathrm{N}$ 在 Y 轴上的坐标范围为 $(a_j, a_{j+1}]$,因此,$q_i\mathrm{NE} \wedge q_j\mathrm{N}$ 在 Y 轴上的坐标范围为

$$(0, a_{i+1}] + (a_j, a_{j+1}] = (a_j, a_{i+1} + a_{j+1}] \tag{7.9}$$

$q_i\mathrm{NE}$ 在 X 轴上的坐标范围为 $(0, a_{i+1}]$,$q_j\mathrm{N}$ 在 X 轴上的坐标范围为 $[0,0]$,因此,$q_i\mathrm{NE} \wedge q_j\mathrm{N}$ 在 X 轴上的坐标范围为

$$(0, a_{i+1}] + [0,0] = (0, a_{i+1}] \tag{7.10}$$

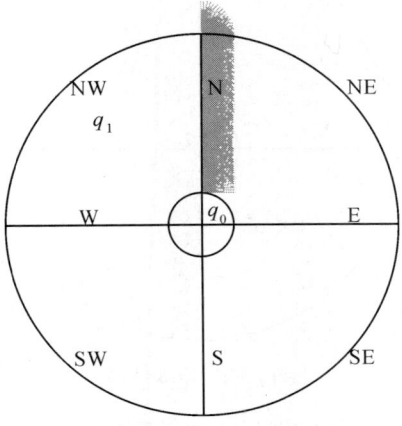

图 7.9 $q_0\mathrm{NE} \wedge q_1\mathrm{N}$ 的模拟结果

根据式(7.8)、式(7.9)和式(7.10)，$q_i\text{NE} \wedge q_j\text{N}$ 的坐标范围为

$$\{(x,y) | x \in (0, a_{i+1}], y \in (a_j, a_{i+1}+a_{j+1}]\} \quad (7.11)$$

根据式(7.11)可得

(1) 当 $i=j$ 时，$\{(x,y) | x \in (0, a_{i+1}], y \in (a_i, a_{i+1}+a_i]\} \in q_i\text{NE} \wedge q_{i+1}\text{NE}$。

(2) 当 $i=j-1$ 时，$\{(x,y) | x \in (0, a_{i+1}], y \in (a_i+a_{i+1}, a_{i+1}+a_{i+2}]\} \in q_j\text{NE} \vee q_{j+1}\text{NE}$。

(3) 当 $i=j+1$ 时，$\{(x,y) | x \in (0, a_{i+1}], y \in (a_{i-1}+a_i, a_{i+1}+a_i]\} \in q_i\text{NE} \vee q_{i+1}\text{NE}$。

当 $i=1, |i-j| \leqslant 1$ 时，$q_1\text{NE} \wedge q_j\text{S}$ 的模拟结果见图 7.10。表 7.4 为 $q_i\text{NE} \wedge q_j\text{S}$ 的推理结果。

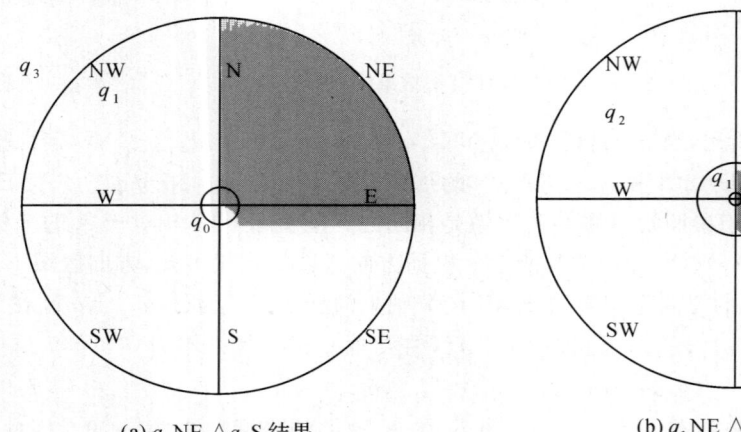

(a) $q_1\text{NE} \wedge q_0\text{S}$ 结果

(b) $q_1\text{NE} \wedge q_1\text{S}$ 结果

(c) $q_1\text{NE} \wedge q_2\text{S}$ 结果

图 7.10 $q_1\text{NE} \wedge q_j\text{S}$ 的模拟结果 $|1-j| \leqslant 1$

表 7.4 $q_i\text{NE} \wedge q_j\text{S}$ 推理组合表

$\wedge$	$q_0\text{S}$	$q_1\text{S}$	$q_2\text{S}$	$q_3\text{S}$	$q_4\text{S}$
$q_0\text{NE}$	$q_0\{\text{NE,E,SE}\}, q_1\text{SE}$	$q_0/q_1\text{SE}$	$q_2\text{S}$	$q_3\text{S}$	$q_4\text{S}$
$q_1\text{NE}$	$q_0/q_1\{\text{NE,E,SE}\}$	$q_0/q_1\{\text{NE,E,SE}\}, q_2\text{SE}$	$q_0/q_1/q_2\text{SE}$	$q_3\text{S}$	$q_4\text{S}$
$q_2\text{NE}$	$q_2\text{NE}$	$q_0\text{NE}, q_1/q_2\{\text{NE,E,SE}\}$	$q_0/q_1/q_2\{\text{NE,E,SE}\}, q_3\text{SE}$	$q_0/q_1/q_2/q_3\text{SE}$	$q_4\text{S}$
$q_3\text{NE}$	$q_3\text{NE}$	$q_3\text{NE}$	$q_0/q_1\text{NE}, q_2/q_3\{\text{NE,E,SE}\}$	$q_0/q_1/q_2/q_3\{\text{NE,E,SE}\}, q_4\text{SE}$	$q_0/q_1/q_2/q_3/q_4\text{SE}$
$q_4\text{NE}$	$q_4\text{NE}$	$q_4\text{NE}$	$q_4\text{NE}$	$q_0/q_1/q_2\text{NE}, q_3/q_4\{\text{NE,E,SE}\}$	$q_0/q_1/q_2/q_3/q_4\{\text{NE,E,SE}\}$

从表 7.4 中的推理结果可以看出:

(1) 当 $i=j+1$ 时, $q_i\text{NE} \wedge q_j\text{S} \rightarrow q_0/\cdots/q_i\text{NE} \vee q_j/q_i\{\text{E,SE}\}$。

(2) 当 $i=j$ 时, $q_i\text{NE} \wedge q_i\text{S} \rightarrow q_0/\cdots/q_i\{\text{NE,E,SE}\} \vee q_{i+1}\text{SE}$。

(3) 当 $i=j-1$ 时, $q_i\text{NE} \wedge q_j\text{S} \rightarrow q_0/\cdots/q_j\text{SE}$。

事实上,根据式(7.8)可得 $q_i\text{NE} \wedge q_j\text{S}$ 的坐标范围为

$$\left\{(x_{AC}, y_{AC}) \middle| \begin{cases} x_{AC} = d(A,B)\cos\theta_A & d(A,B) \in (a_i, a_{i+1}] \\ & d(B,C) \in (a_j, a_{j+1}] \\ y_{AC} = d(A,B)\sin\theta_A - d(B,C) & 0 < \theta_A < \frac{\pi}{2} \end{cases}\right\} \quad (7.12)$$

由式(7.12)可得

$$y_{AC} \in (a_i\sin\theta_A - a_{j+1}, a_{i+1}\sin\theta_A - a_j) \quad (0 < \theta_A < \frac{\pi}{2}) \tag{7.13}$$

根据式(7.13)可得

(1) 当 $i=j+1$ 时, $y_{AC} \in (a_i\sin\theta_A - a_i, a_{i+1}\sin\theta_A - a_{i-1})$,再由式(7.12)得

$$q_i\text{NE} \wedge q_{i-1}\text{S} \rightarrow q_0/\cdots/q_i\text{NE} \vee q_j/q_i\{\text{E,SE}\}$$

(2) 当 $j=i$ 时, $y_{AC} \in (a_i\sin\theta_A - a_{i+1}, a_{i+1}\sin\theta_A - a_i)$,再由式(7.12)得

$$q_i\text{NE} \wedge q_i\text{S} \rightarrow q_0/\cdots/q_i\{\text{NE,E,SE}\} \vee q_{i+1}\text{SE}$$

(3) 当 $i=j-1$ 时, $y_{AC} \in (a_i\sin\theta_A - a_{i+2}, a_{i+1}\sin\theta_A - a_{i+1}) \subseteq (-a_{i+2}, 0)$,再由式(7.12)得

$$q_i\text{NE} \wedge q_{i+1}\text{S} \rightarrow q_0/\cdots/q_j\text{SE}$$

由于方向区域 SE 和 NW、N 和 E、W 和 S 关于方向区域 NE 对称,将上述推理结果分别转为其对称部分即可得到 $q_i\text{NE} \wedge q_j\text{NW}$、$q_i\text{NE} \wedge q_j\text{E}$ 以及 $q_i\text{NE} \wedge q_j\text{W}$ 的推理结果。

§7.2 三维位置关系的定性描述及其定性推理

7.2.1 位置关系定性描述

用方向关系和定性距离描述三维空间中空间对象位置关系的方法和描述二维空间中空间对象位置关系的方法类似,采用定性距离在前、方向关系在后的方法,即 $d_{AB}dir(A,B)$,其中 A、B 为空间对象,定性距离 d_{AB} 采用 5 个定性距离:q_0、q_1、q_2、q_3、q_4,如图 7.11 所示。采用投影法将参照物所在空间划分为 27 个方向区域 $O_{i,j}$,$i \in \{N, NE, E, SE, S, SW, W, NW, same\}$,$j \in \{up, between, down\}$,建立方向关系描述模型。定性距离和方向关系划分后的三维空间如图 7.12 所示(刘新 等,2008;Liu Xin et al,2008b)。

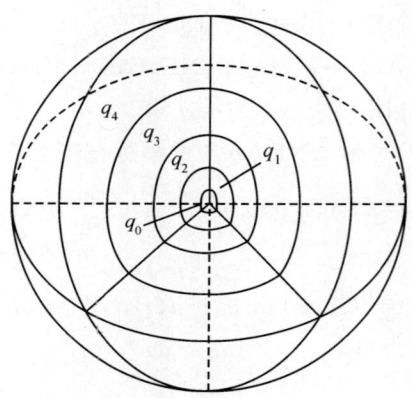

图 7.11 被定性距离划分的三维空间

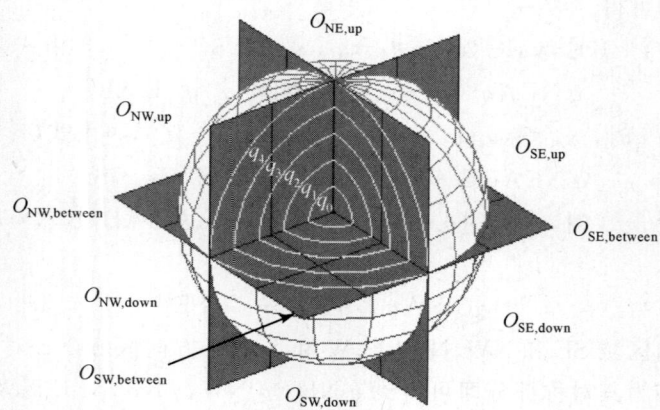

图 7.12 被定性距离和方向关系划分的三维空间

7.2.2 位置关系定性推理

在三维空间中，用定性距离与方向关系描述目标对象与参照物之间的相对位置关系，如 $q_iO_{k,l}$, $q_jO_{k,l}$ 等，其中，方向区域 $O_{k,l}$, $k\in\{\text{N,NE,E,SE,S,SW,W,NW,same}\}$, $l\in\{\text{up,between,down}\}$。用角度描述 $O_{k,l}$ 为 $k\in\{\frac{\pi}{2},(0,\frac{\pi}{2}),0,(\frac{3\pi}{2},2\pi),\frac{3\pi}{2},(\pi,\frac{3\pi}{2}),\pi,(\frac{\pi}{2},\pi)\}$, $l\in\{[0,\frac{\pi}{2}),\frac{\pi}{2},(\frac{\pi}{2},\pi]\}$，例如，用角度描述方向区域 $O_{\text{NE,up}}$ 为 $\theta_A\in(0,\frac{\pi}{2})$, $\beta_A\in[0,\frac{\pi}{2})$。

下面以 $q_iO_{k,l}$ 为例，说明 $q_iO_{k,l}$ 的区间范围。

$$q_iO_{k,l}=\{(x_B,y_B,z_B)\in O_{k,j}|\Delta_{i-1}\leqslant\sqrt{(x_A-x_B)^2+(y_A-y_B)^2+(z_A-z_B)^2}\leqslant\Delta_{i-1}+\delta_i\} \tag{7.14}$$

用球坐标表示 $q_iO_{k,l}$ 为

$$q_iO_{k,l}=\{(x_B,y_B,z_B)\in O_{i,j}|\Delta_{i-1}\leqslant r_{AB}\leqslant\Delta_{i-1}+\delta_i\} \tag{7.15}$$

球坐标与直角坐标之间的变换公式为

$$\begin{cases} x=r\sin\beta\cos\theta & 0\leqslant r<+\infty \\ y=r\sin\beta\sin\theta & 0\leqslant\beta\leqslant\pi \\ z=r\cos\beta & 0\leqslant\theta<2\pi \end{cases} \tag{7.16}$$

式中，$r=$constant 是以原点为球心的球面，$\beta=$constant 是以原点为顶点、Z 轴为中心轴的圆锥面，$\theta=$constant 是过 Z 轴的半平面（吕林根 等，1990）。

在三维球面坐标系中，设点状物体 A、B、C 的坐标分别为 $A(x_A,y_A,z_A)$, $B(x_B,y_B,z_B)$, $C(x_C,y_C,z_C)$, 且 $d(A,B)\in(\Delta_{i-1},\Delta_{i-1}+\delta_i]$, $d(B,C)\in(\Delta_{j-1},\Delta_{j-1}+\delta_j]$。当 B 和参照物 A 的距离 $d(A,B)\leqslant\Delta_{i-1}+\delta_i$ 时，B 的范围为

$$\left\{(x_B,y_B,z_B)\middle| \begin{cases} x_B-x_A=d(A,B)\sin\beta_A\cos\theta_A & \beta_A\in[0,\pi] \\ y_B-y_A=d(A,B)\sin\beta_A\sin\theta_A & \theta_A\in[0,2\pi) \\ z_B-z_A=d(A,B)\cos\beta_A \end{cases}\right\} \tag{7.17}$$

根据式(7.17)，当 B 和参照物 A 的距离 $d(A,B)\in(\Delta_i,\Delta_{i-1}+\delta_i]$ 时，B 的范围为

$$\left\{(x_B,y_B,z_B)\middle| \begin{cases} x_B-x_A=d(A,B)\sin\beta_A\cos\theta_A & d(A,B)\leqslant\Delta_{i-1}+\delta_i \\ y_B-y_A=d(A,B)\sin\beta_A\sin\theta_A & \beta_A\in[0,\pi] \\ z_B-z_A=d(A,B)\cos\beta_A & \theta_A\in[0,2\pi) \end{cases}\right\} -$$

$$\left\{(x_B,y_B,z_B)\middle| \begin{cases} x_B-x_A=d(A,B)\sin\beta_A\cos\theta_A & d(A,B)\leqslant\Delta_{i-1} \\ y_B-y_A=d(A,B)\sin\beta_A\sin\theta_A & \beta_A\in[0,\pi] \\ z_B-z_A=d(A,B)\cos\beta_A & \theta_A\in[0,2\pi) \end{cases}\right\}$$

$$= \left\{ (x_B, y_B, z_B) \,\middle|\, \begin{cases} x_B - x_A = d(A,B)\sin\beta_A\cos\theta_A & d(A,B) \in (\Delta_{i-1}, \Delta_{i-1}+\delta_i] \\ y_B - y_A = d(A,B)\sin\beta_A\sin\theta_A & \beta_A \in [0,\pi] \\ z_B - z_A = d(A,B)\cos\beta_A & \theta_A \in [0,2\pi] \end{cases} \right\} \quad (7.18)$$

同理可得，当 C 和参照物 B 的距离 $d(B,C) \in (\Delta_{j-1}, \Delta_{j-1}+\delta_j]$ 时，C 的范围为

$$\left\{ (x_C, y_C, z_C) \,\middle|\, \begin{cases} x_C - x_B = d(B,C)\sin\beta_B\cos\theta_B & d(B,C) \in (\Delta_{j-1}, \Delta_{j-1}+\delta_j] \\ y_C - y_B = d(B,C)\sin\beta_B\sin\theta_B & \beta_B \in [0,\pi] \\ z_C - z_B = d(B,C)\cos\beta_B & \theta_B \in [0,2\pi] \end{cases} \right\} \quad (7.19)$$

当 $d(A,B) \in (\Delta_{i-1}, \Delta_{i-1}+\delta_i]$，且 $d(B,C) \in (\Delta_{j-1}, \Delta_{j-1}+\delta_j]$ 时，C 的范围为

$$\left\{ (x_B, y_B, z_B) \,\middle|\, \begin{cases} x_B - x_A = d(A,B)\sin\beta_A\cos\theta_A & d(A,B) \in (\Delta_{i-1}, \Delta_{i-1}+\delta_i] \\ y_B - y_A = d(A,B)\sin\beta_A\sin\theta_A & \beta_A \in [0,\pi] \\ z_B - z_A = d(A,B)\cos\beta_A & \theta_A \in [0,2\pi] \end{cases} \right\} \cap$$

$$\left\{ (x_C, y_C, z_C) \,\middle|\, \begin{cases} x_C - x_B = d(B,C)\sin\beta_B\cos\theta_B & d(B,C) \in (\Delta_{j-1}, \Delta_{j-1}+\delta_j] \\ y_C - y_B = d(B,C)\sin\beta_B\sin\theta_B & \beta_B \in [0,\pi] \\ z_C - z_B = d(B,C)\cos\beta_B & \theta_B \in [0,2\pi] \end{cases} \right\}$$

$$= \left\{ (x_C, y_C, z_C) \,\middle|\, \begin{cases} x_C - x_A = d(A,B)\sin\beta_A\cos\theta_A + d(B,C)\sin\beta_B\cos\theta_B \\ y_C - y_A = d(A,B)\sin\beta_A\sin\theta_A + d(B,C)\sin\beta_B\sin\theta_B & \beta_A, \beta_B \in [0,\pi] \\ z_C - z_A = d(A,B)\cos\beta_A + d(B,C)\cos\beta_B & \theta_A, \theta_B \in [0,2\pi] \end{cases} \right\}$$
$$(7.20)$$

令 $x_{AC} = x_C - x_A, y_{AC} = y_C - y_A, z_{AC} = z_C - z_A$，式(7.20)写为

$$\left\{ (x_{AC}, y_{AC}, z_{AC}) \,\middle|\, \begin{cases} x_{AC} = d(A,B)\sin\beta_A\cos\theta_A + d(B,C)\sin\beta_B\cos\theta_B \\ y_{AC} = d(A,B)\sin\beta_A\sin\theta_A + d(B,C)\sin\beta_B\sin\theta_B & \beta_A, \beta_B \in [0,\pi] \\ z_{AC} = d(A,B)\cos\beta_A + d(B,C)\cos\beta_B & \theta_A, \theta_B \in [0,2\pi] \end{cases} \right\}$$
$$(7.21)$$

式(7.21)给出了根据 B 相对于 A 的位置关系和 C 相对于 B 的位置关系，求得 C 相对于 A 的位置关系的表达式，如 $q_i O_{\text{NE,up}} \wedge q_j O_{\text{SE,up}}$ 的坐标范围

$$\left\{ (x_{AC}, y_{AC}, z_{AC}) \,\middle|\, \begin{cases} x_{AC} = d(A,B)\sin\beta_A\cos\theta_A + & d(A,B) \in (A_i, A_{i+1}] \\ \quad d(B,C)\sin\beta_B\cos\theta_B & d(B,C) \in (A_j, A_{j+1}] \\ y_{AC} = d(A,B)\sin\beta_A\sin\theta_A + & \beta_A, \beta_B \in [0, \frac{\pi}{2}) \\ \quad d(B,C)\sin\beta_B\sin\theta_B & \theta_A \in (0, \frac{\pi}{2}) \\ z_{AC} = d(A,B)\cos\beta_A + d(B,C)\cos\beta_B & \theta_B \in (\frac{3\pi}{2}, 2\pi) \end{cases} \right\} \quad (7.22)$$

设定性距离相邻距离区间比为 $k=5$，并假设当 $|i-j| \geqslant 2$ 时，满足 $q_i \gg q_j$ 或

$q_j \gg q_i$,根据式(7.21)进行推理,推理后的部分结果见表 7.5 至表 7.8。

表 7.5 $q_i O_{\text{NE,up}} \wedge q_j O_{\text{SE,up}}$ 推理组合

$\wedge$	$q_0 O_{\text{SE,up}}$	$q_1 O_{\text{SE,up}}$	$q_2 O_{\text{SE,up}}$	$q_3 O_{\text{SE,up}}$	$q_4 O_{\text{SE,up}}$
$q_0 O_{\text{NE,up}}$	$q_0/q_1 D_{\text{E-up}}$	$q_0/q_1/q_2 D_{\text{E-up}}$	$q_2 O_{\text{SE,up}}$	$q_3 O_{\text{SE,up}}$	$q_4 O_{\text{SE,up}}$
$q_1 O_{\text{NE,up}}$	$q_0/q_1/q_2 D_{\text{E-up}}$	$q_0/q_1/q_2 D_{\text{E-up}}$	$q_0/q_1/q_2/q_3 D_{\text{E-up}}$	$q_3 O_{\text{SE,up}}$	$q_4 O_{\text{SE,up}}$
$q_2 O_{\text{NE,up}}$	$q_2 O_{\text{NE,up}}$	$q_0 O_{\text{NE,up}}$, $q_1/q_2/q_3 D_{\text{E-up}}$	$q_0/q_1/q_2/q_3 D_{\text{E-up}}$	$q_0 O_{\text{SE,up}}$, $q_1/q_2/q_3/q_4 D_{\text{E-up}}$	$q_4 O_{\text{SE,up}}$
$q_3 O_{\text{NE,up}}$	$q_3 O_{\text{NE,up}}$	$q_3 O_{\text{NE,up}}$	$q_0/q_1 O_{\text{NE,up}}$, $q_2/q_3/q_4 D_{\text{E-up}}$	$q_0/q_1/q_2/q_3/q_4 D_{\text{E-up}}$	$q_0/q_1 O_{\text{NE,up}}$, $q_2/q_3/q_4 D_{\text{E-up}}$
$q_4 O_{\text{NE,up}}$	$q_4 O_{\text{NE,up}}$	$q_4 O_{\text{NE,up}}$	$q_4 O_{\text{NE,up}}$	$q_0/q_1/q_2 O_{\text{NE,up}}$, $q_3/q_4 D_{\text{E-up}}$	$q_0/q_1/q_2/q_3/q_4 D_{\text{E-up}}$

注:$D_{\text{E-up}} = \{O_{\text{NE,up}}, O_{\text{E,up}}, O_{\text{SE,up}}\}$。

图 7.13 是根据式(7.21)模拟 $q_1 O_{\text{NE,up}} \wedge q_j O_{\text{SE,up}}$ 推理的结果,其中 $|1-j| \leqslant 1$,为了便于观察只画出相关部分。

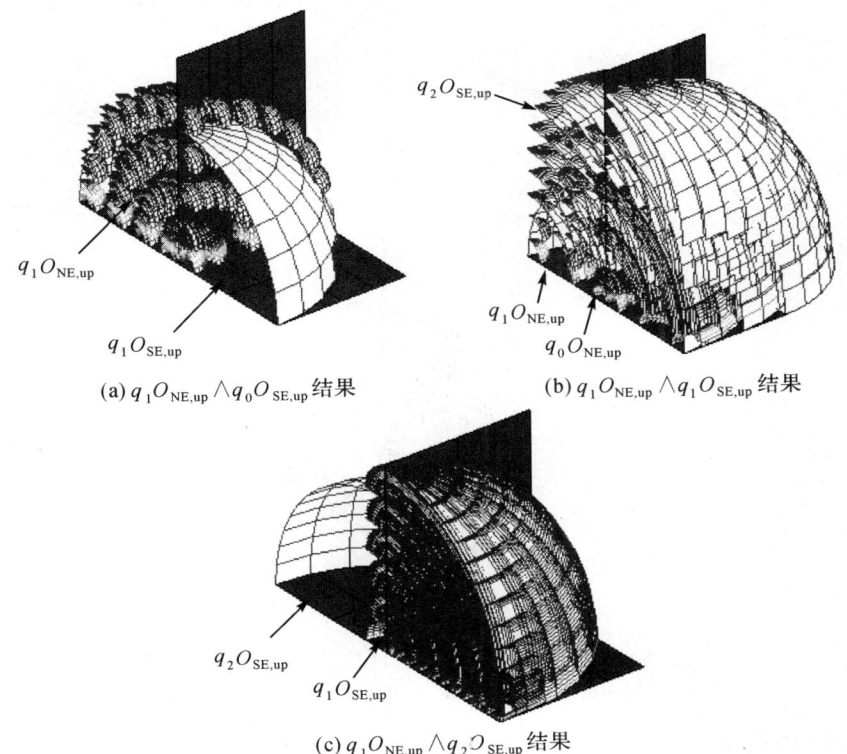

图 7.13 $q_1 O_{\text{NE,up}} \wedge q_j O_{\text{SE,up}}$ 的模拟结果($|1-j| \leqslant 1$)

根据式(7.21)，$q_iO_{NE,up} \wedge q_jO_{SE,down}$ 推理结果为表 7.6，图 7.14 是模拟 $q_1O_{NE,up} \wedge q_jO_{SE,down}$ 推理的结果，其中 $|1-j|\leqslant 1$。

表 7.6 　$q_iO_{NE,up} \wedge q_jO_{SE,down}$ 推理组合

$\wedge$	$q_0O_{SE,down}$	$q_1O_{SE,down}$	$q_2O_{SE,down}$	$q_3O_{SE,down}$	$q_4O_{SE,down}$
$q_0O_{NE,up}$	$q_0/q_1 D_E$	$q_0/q_1/q_2 D_E$	$q_2O_{SE,down}$	$q_3O_{SE,down}$	$q_4O_{SE,down}$
$q_1O_{NE,up}$	$q_0/q_1/q_2 D_E$	$q_0/q_1/q_2 D_E$	$q_0/q_1/q_2/q_3 D_E$	$q_3O_{SE,down}$	$q_4O_{SE,down}$
$q_2O_{NE,up}$	$q_2O_{NE,up}$	$q_0/q_1/q_2/q_3 D_E$	$q_0/q_1/q_2/q_3 D_E$	$q_0/q_1/q_2/q_3/q_4 D_E$	$q_4O_{SE,down}$
$q_3O_{NE,up}$	$q_3O_{NE,up}$	$q_3O_{NE,up}$	$q_0/q_1/q_2/q_3/q_4 D_E$	$q_0/q_1/q_2/q_3/q_4 D_E$	$q_0/q_1/q_2/q_3/q_4 D_E$
$q_4O_{NE,up}$	$q_4O_{NE,up}$	$q_4O_{NE,up}$	$q_4O_{NE,up}$	$q_0/q_1/q_2/q_3/q_4 D_E$	$q_0/q_1/q_2/q_3/q_4 D_E$

注：$D_E=\{O_{NE,j},O_{E,j},O_{SE,j}\}, j\in\{up,between,down\}$。

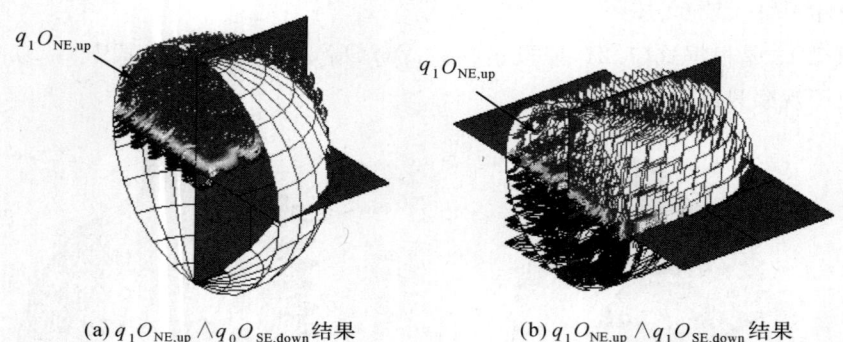

(a) $q_1O_{NE,up} \wedge q_0O_{SE,down}$ 结果　　(b) $q_1O_{NE,up} \wedge q_1O_{SE,down}$ 结果

(c) $q_1O_{NE,up} \wedge q_2O_{SE,down}$ 结果

图 7.14 　$q_1O_{NE,up} \wedge q_jO_{SE,down}$ 的模拟结果($1-j|\leqslant 1$)

根据式(7.21)，$q_iO_{NE,up} \wedge q_jO_{SE,down}$ 推理结果为表 7.7，图 7.15 是 $q_1O_{SE,up} \wedge q_0O_{NW,up}$ 的结果，图 7.15(a)是没有方向关系的划分平面，图 7.15(b)是图 7.15(a)

第 7 章 位置信息的定性描述及其定性推理

加上方向关系划分平面后的结果，图 7.15(c) 是图 7.15(b) 的结果在平面 $z=z_A$ 上的投影图。图 7.16 是 $q_1O_{SE,up} \wedge q_1O_{NW,up}$ 的结果，其中，图 7.16(a) 是没有方向关系的划分平面，图 7.16(b) 是图 7.16(a) 加上方向关系划分平面后的结果，图 7.16(c) 是图 7.16(b) 的结果在平面 $z=z_A$ 上的投影图。图 7.17 是 $q_1O_{SE,up} \wedge q_2O_{NW,up}$ 的结果，其中，图 7.17(a) 是没有方向关系的划分平面，图 7.17(b) 是图 7.17(a) 加上方向关系划分平面后的结果，图 7.17(c) 是图 7.17(b) 的结果在平面 $z=z_A$ 上的投影图。

表 7.7　$q_iO_{SE,up} \wedge q_jO_{NW,up}$ 推理组合

$\wedge$	$q_0O_{NW,up}$	$q_1O_{NW,up}$	$q_2O_{NW,up}$	$q_3O_{NW,up}$	$q_4O_{NW,up}$
$q_0O_{SE,up}$	$q_0/q_1 D_{up}$	$q_0/q_1/q_2 D_{up}$	$q_2O_{NW,up}$	$q_3O_{NW,up}$	$q_4O_{NW,up}$
$q_1O_{SE,up}$	$q_0/q_1/q_2 D_{up}$	$q_0/q_1/q_2 D_{up}$	$q_0/q_1/q_2/q_3 D_{up}$	$q_3O_{NW,up}$	$q_4O_{NW,up}$
$q_2O_{SE,up}$	$q_2O_{SE,up}$	$q_0/q_1/q_2/q_3 D_{up}$	$q_0/q_1/q_2/q_3 D_{up}$	$q_0/q_1/q_2/q_3/q_4 D_{up}$	$q_4O_{NW,up}$
$q_3O_{SE,up}$	$q_3O_{SE,up}$	$q_3O_{SE,up}$	$q_0/q_1/q_2/q_3/q_4 D_{up}$	$q_0/q_1/q_2/q_3/q_4 D_{up}$	$q_0/q_1/q_2/q_3/q_4 D_{up}$
$q_4O_{SE,up}$	$q_4O_{SE,up}$	$q_4O_{SE,up}$	$q_4O_{SE,up}$	$q_0/q_1/q_2/q_3/q_4 D_{up}$	$q_0/q_1/q_2/q_3/q_4 D_{up}$

注：$D_{up} = \{O_{NE,up}, O_{E,up}, O_{SE,up}, O_{S,up}, O_{SW,up}, O_{W,up}, O_{NW,up}, O_{N,up}, O_{same,up}\}$。

根据表 7.7 可知，对任意的 $i \in \{0,1,2,3,4\}$，$q_iO_{SE,up} \wedge q_iO_{NW,up}$，$q_iO_{SE,up} \wedge q_{i+1}O_{NW,up}$ 和 $q_{i+1}O_{SE,up} \wedge q_iO_{NW,up}$ 推理结果相同，均为 $q_0/q_1/\cdots/q_{\max(i,j)}\{O_{NE,up}, O_{E,up}, O_{SE,up}, O_{S,up}, O_{SW,up}, O_{W,up}, O_{NW,up}, O_{N,up}, O_{same,up}\}$。事实上，$x_{AC}$、$y_{AC}$ 的范围是截然不同的。图 7.15(c) 给出了 $i=1$ 时，$q_1O_{SE,up} \wedge q_0O_{NW,up}$ 在 $z=z_A$ 上的投影图，图 7.16(c) 是 $i=1$ 时，$q_1O_{SE,up} \wedge q_1O_{NW,up}$ 在 $z=z_A$ 上的投影图，图 7.18 是 $i=1$ 时，$q_0O_{SE,up} \wedge q_1O_{NW,up}$ 在 $z=z_A$ 上的投影图。对于其他的 i 值，$q_{i+1}O_{SE,up} \wedge q_iO_{NW,up}$ 在 $z=z_A$ 上的投影图与图 7.15(c) 相似，$q_iO_{SE,up} \wedge q_iO_{NW,up}$ 在 $z=z_A$ 上的投影图与图 7.16(c) 相似，$q_iO_{SE,up} \wedge q_{i+1}O_{NW,up}$ 在 $z=z_A$ 上的投影图与图 7.18 相似，如图 7.17(c) 与图 7.18(c) 相似。

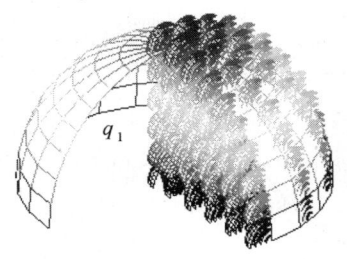 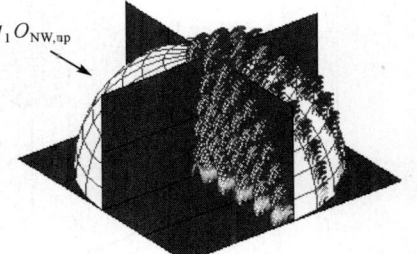

(a) $q_1O_{SE,up} \wedge q_0O_{NW,up}$ 在定性距离系统中的推理结果　　(b) $q_1O_{SE,up} \wedge q_0O_{NW,up}$ 结果

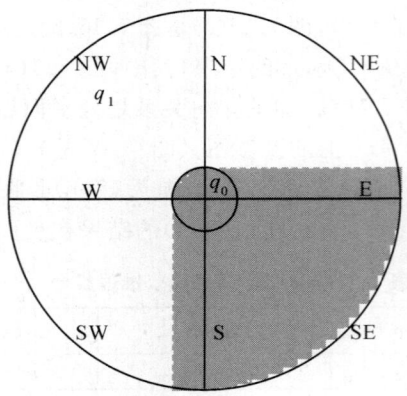

(c) $q_1 O_{SE,up} \wedge q_0 O_{NW,up}$ 在 $z=z_A$ 上的投影

图 7.15　$q_1 O_{SE,up} \wedge q_0 O_{NW,up}$ 结果

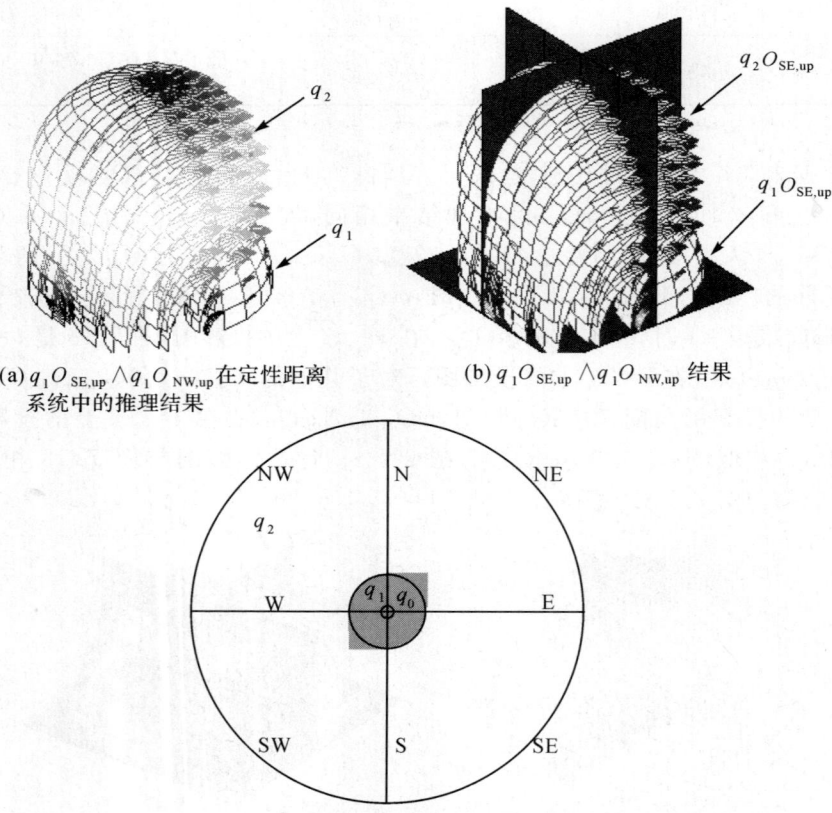

(a) $q_1 O_{SE,up} \wedge q_1 O_{NW,up}$ 在定性距离系统中的推理结果

(b) $q_1 O_{SE,up} \wedge q_1 O_{NW,up}$ 结果

(c) $q_1 O_{SE,up} \wedge q_1 O_{NW,up}$ 在 $z=z_A$ 上的投影

图 7.16　$q_1 O_{SE,up} \wedge q_1 O_{NW,up}$ 结果

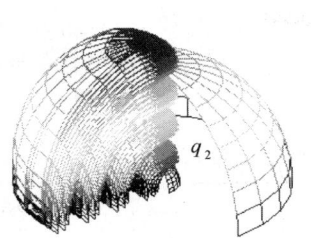
(a) $q_1O_{SE,up} \wedge q_2O_{NW,up}$ 在定性距离系统中的推理结果

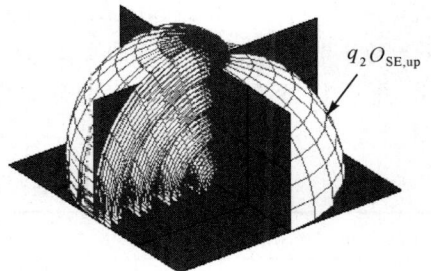
(b) $q_1O_{SE,up} \wedge q_2O_{NW,up}$ 结果

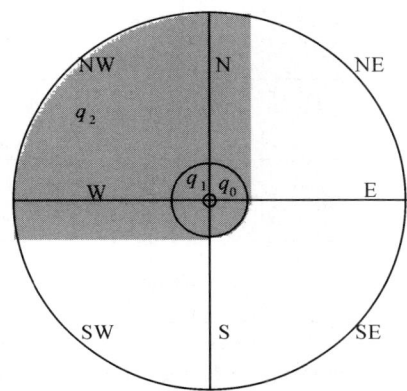
(c) $q_1O_{SE,up} \wedge q_2O_{NW,up}$ 在 $z=z_A$ 上的投影

图 7.17　$q_1O_{SE,up} \wedge q_2O_{NW,up}$ 结果

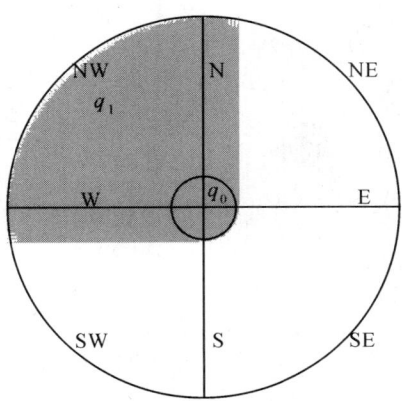
图 7.18　$q_0O_{SE,up} \wedge q_1O_{NW,up}$ 在 $z=z_A$ 上的投影

表7.8是根据式(7.21),$q_iO_{\text{SE,up}} \wedge q_jO_{\text{NW,down}}$的推理结果,图7.19至图7.21中的(a)分图分别为 $q_1O_{\text{SE,up}} \wedge q_0O_{\text{NW,down}}$, $q_1O_{\text{SE,up}} \wedge q_1O_{\text{NW,down}}$ 和 $q_1O_{\text{SE,up}} \wedge q_2O_{\text{NW,down}}$ 的结果;图7.19至图7.21中的(b)分图分别是(a)分图的推理结果在 $z=z_A$ 上的投影图。

表7.8　$q_iO_{\text{SE,up}} \wedge q_jO_{\text{NW,down}}$ 推理组合

$\wedge$	$q_0O_{\text{NW,down}}$	$q_1O_{\text{NW,down}}$	$q_2O_{\text{NW,down}}$	$q_3O_{\text{NW,down}}$	$q_4O_{\text{NW,down}}$
$q_0O_{\text{SE,up}}$	q_0D, q_1D_3	q_0D, q_1D_1	$q_2O_{\text{NW,down}}$	$q_3O_{\text{NW,down}}$	$q_4O_{\text{NW,down}}$
$q_1O_{\text{SE,up}}$	q_0D q_1D_2	q_0/q_1D, q_2D_3	q_0/q_1D, q_2D_1	$q_3O_{\text{NW,down}}$	$q_4O_{\text{NW,down}}$
$q_2O_{\text{SE,up}}$	$q_2O_{\text{SE,up}}$	q_0/q_1D, q_2D_2	$q_0/q_1/q_2D$, q_3D_3	$q_0/q_1/q_2D$, q_3D_1	$q_4O_{\text{NW,down}}$
$q_3O_{\text{SE,up}}$	$q_3O_{\text{SE,up}}$	$q_3O_{\text{SE,up}}$	$q_0/q_1/q_2D$, q_3D_2	$q_0/q_1/q_2/q_3D$, q_4D_3	$q_0/q_1/q_2/q_3D$, q_4D_1
$q_4O_{\text{SE,up}}$	$q_4O_{\text{SE,up}}$	$q_4O_{\text{SE,up}}$	$q_4O_{\text{SE,up}}$	$q_0/q_1/q_2/q_3D$, q_4D_2	$q_0/q_1/q_2/q_3/$ q_4D

注:$D=\{O_{\text{NE},j}, O_{\text{E},j}, O_{\text{SE},j}, O_{\text{S},j}, O_{\text{SW},j}, O_{\text{W},j}, O_{\text{NW},j}, O_{\text{N},j}, O_{\text{same},j}\}, j \in \{\text{up, between, down}\}$;
$D_1=\{O_{\text{NE},j}, O_{\text{N},j}, O_{\text{NW},j}, O_{\text{W},j}, O_{\text{SW},j}, O_{\text{S,down}}, O_{\text{SE,down}}, O_{\text{E,down}}, O_{\text{same,down}}\}$;
$D_2=\{O_{\text{NE},j}, O_{\text{E},j}, O_{\text{SE},j}, O_{\text{S},j}, O_{\text{SW},j}, O_{\text{W,up}}, O_{\text{NW,up}}, O_{\text{N,up}}, O_{\text{same,up}}\}$;
$D_3=\{O_{\text{SW},j}, O_{\text{W,up}}, O_{\text{NE},j}, O_{\text{N,up}}, O_{\text{NW,up}}, O_{\text{SE,down}}\}$。

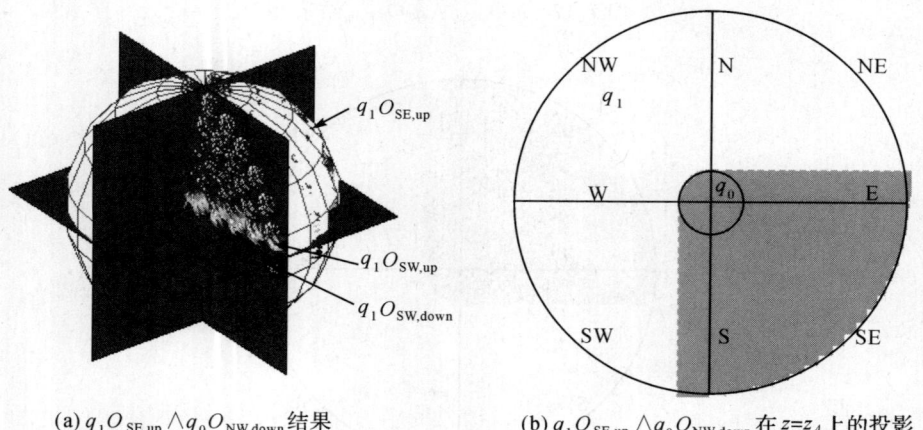

(a) $q_1O_{\text{SE,up}} \wedge q_0O_{\text{NW,down}}$ 结果　　(b) $q_1O_{\text{SE,up}} \wedge q_0O_{\text{NW,down}}$ 在 $z=z_A$ 上的投影

图7.19　$q_1O_{\text{SE,up}} \wedge q_0O_{\text{NW,down}}$ 结果

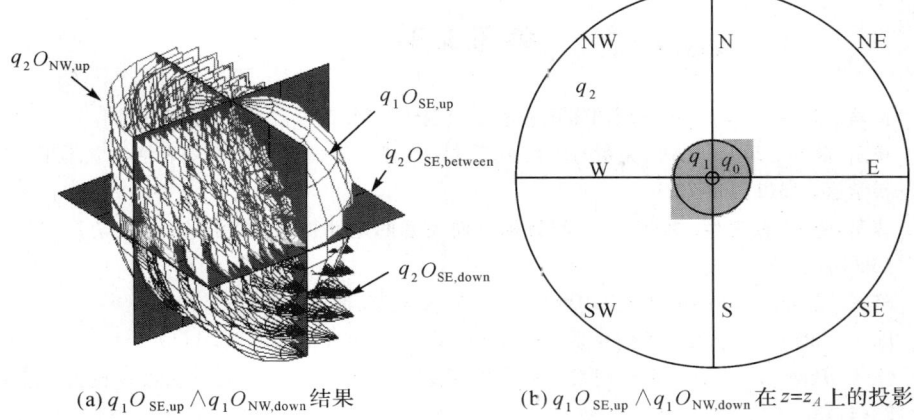

(a) $q_1O_{SE,up} \wedge q_1O_{NW,down}$ 结果　　(b) $q_1O_{SE,up} \wedge q_1O_{NW,down}$ 在 $z=z_A$ 上的投影

图 7.20　$q_1O_{SE,up} \wedge q_1O_{NW,down}$ 结果

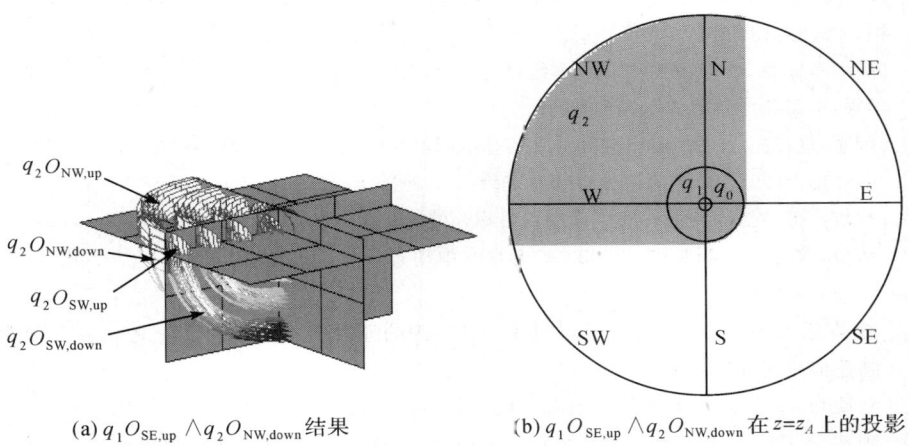

(a) $q_1O_{SE,up} \wedge q_2O_{NW,down}$ 结果　　(b) $q_1O_{SE,up} \wedge q_2O_{NW,down}$ 在 $z=z_A$ 上的投影

图 7.21　$q_1O_{SE,up} \wedge q_2O_{NW,down}$ 结果

参考文献

[1] 曹菡. 2002. 空间关系推理的知识表示与推理机制研究[D]. 武汉:武汉大学.

[2] 曹菡,陈军. 2001a. 方向关系与距离关系的定性描述与推理[J]. 西安石油学院学报:自然科学版,16(1):68-72.

[3] 曹菡,陈军,杜道生. 2001b. 空间目标方向关系的定性扩展描述[J]. 测绘学报,30(2):162-167.

[4] 陈军. 2002. Voronoi 动态空间数据模型[M]. 北京:测绘出版社.

[5] 陈军. 2003. 论中国地理信息系统的发展方向[J]. 地理信息世界,1(1):6-11.

[6] 陈军,郭薇. 1998a. 三维空间实体间拓扑关系的矩阵描述[J]. 武汉测绘科技大学学报,23(4):359-363.

[7] 陈军,郭薇. 1998b. 基于剖分的三维拓扑 ER 模型研究[J]. 测绘学报,27(4):308-317.

[8] 陈军,蒋捷. 2000. 多维动态 GIS 的空间数据建模、处理与分析[J]. 武汉测绘科技大学学报,25(3):189-195.

[9] 陈军,李志林,蒋捷,等. 2004. 多维动态 GIS 空间数据模型与方法的研究[J]. 武汉大学学报:信息科学版,29(10):858-862.

[10] 陈军,赵仁亮. 1999. GIS 空间关系的基本问题与研究进展[J]. 测绘学报,28(2):95-102.

[11] 陈述彭. 1997. 遥感地学分析的时空维[J]. 遥感学报,1(3):161-171.

[12] 邓敏,刘文宝. 1999. GIS 中模糊点目标位置的不确定性描述[J]. 矿山测量,(3):13-15.

[13] 邓敏,刘文宝,崔先国. 2002. 矿山 GIS 中动态点位置的不确定性表达[J]. 煤炭学报,27(2):152-157.

[14] 杜世宏. 2004. 空间关系模糊描述及组合推理的理论和方法研究[D]. 北京:中国科学院遥感应用研究所.

[15] 杜晓初. 2005. 多重表达中空间拓扑关系等价性研究[D]. 武汉:武汉大学.

[16] 郭达志. 2002. 地理信息系统原理与应用[M]. 徐州:中国矿业大学出版社.

[17] 郭平. 2004. 定性空间推理技术及应用研究[D]. 重庆:重庆大学.

[18] 郭仁忠. 1997. 空间分析[M]. 武汉:武汉测绘科技大学出版社.

[19] 郭薇. 1998. 顾及空间剖分的三维拓扑空间数据模型[D]. 武汉:武汉测绘科技大学.

[20] 郭薇,陈军. 1997a. 基于流形拓扑的三维空间实体形式化描述[J]. 武汉测绘科技大学学报,27(3):156-161.

[21] 郭薇,陈军. 1997b. 基于点集拓扑学的三维拓扑空间关系形式化描述[J]. 测绘学报,26(2):122-127.

[22] 何建华,刘耀林. 2004. GIS 中拓扑和方向关系推理模型[J]. 测绘学报,33(2):156-162.

[23] 李成名. 1998. 基于 Voronoi 图的空间关系描述、表达与推断[D]. 武汉:武汉测绘科技大学.

[24] 李德仁. 1997a. 论 RS、GPS 与 GIS 集成的定义、理论与关键技术[J]. 遥感学报,1(1):64-68.

[25] 李德仁. 1997b. 关于地理信息系统理论的若干思考[J]. 武汉测绘科技大学学报, 22(2):93-95.

[26] 李德仁,龚健雅,边馥苓. 1993. 地理信息系统导论[M]. 北京:测绘出版社.

[27] 李清泉. 1998. 基于混合结构的三维GIS数据模型与空间分析研究[D]. 武汉:武汉测绘科技大学.

[28] 梁启章. 1995. GIS和计算机制图[M]. 北京:科学出版社.

[29] 刘炳初. 2001. 泛函分析[M]. 北京:科学出版社.

[30] 刘文宝,邓敏,夏宗国. 2000. 矢量GIS中属性数据的不确定性分析[J]. 测绘学报, 29(1):76-81.

[31] 刘新,刘文宝. 2006. 3D-GIS中方向关系定性推理研究[J]. 辽宁工程技术大学学报:自然科学版, 25(sup):39-41.

[32] 刘新,刘文宝. 2007a. 空间数据挖掘中关联规则的支持度和可信度研究[J]. 测绘科学技术学报, 24(2):93-96.

[33] 刘新,刘文宝. 2007b. 3D-GIS中方向关系描述及其推理[J]. 测绘科学, 32(3):23-25.

[34] 刘新,刘文宝,李成名. 2008. 三维GIS中位置关系的定性描述与推理[J]. 测绘学报, 37(4):495-500.

[35] 吕林根,许子道. 1987. 解析几何[M]. 3版. 北京:高等教育出版社.

[36] 孙海滨. 2006. 定性空间推理及其应用技术研究[D]. 长春:吉林大学.

[37] 孙玉国. 1993. 拓扑关系描述与2D T-String空间关系表达[D]. 武汉:武汉测绘科技大学.

[38] 万剑华. 2001. 城市三维地理信息系统的建模研究[D]. 武汉:武汉大学.

[39] 吴立新. 2004. 真3维地学构模的若干问题[J]. 地理信息世界, 2(3):13-18.

[40] 吴立新,史文中. 2005. 论三维地学空间构模[J]. 地理与地理信息科学, 21(1):1-4.

[41] 谢琦. 2006. 空间方位关系模型与时空结合推理的研究[D]. 长春:吉林大学.

[42] 熊金城. 1981. 点集拓扑学讲义[M]. 北京:高等教育出版社.

[43] 易善桢. 1999. 基于单纯形的3D-GIS数据模型及其初步设计[J]. 测绘通报, (11):10-13.

[44] 张文修. 1984. 模糊数学基础[M]. 西安:西安交通大学出版社.

[45] 赵红超. 2006. 空间关系的研究和实现[D]. 北京:中国科学院计算技术研究所.

[46] 赵仁亮. 2002. 基于Voronoi图的空间关系计算研究[D]. 长沙:中南大学.

[47] 朱铁稳. 2002. 基于均匀空间离散域对象的空间数据库关键技术研究[D]. 长沙:国防科学技术大学.

[48] ABDELMOTY A I, EI-GERESY B A. 1995. A General Approach to the Representation of Spatial Relationships:Technical Report CS-95-6[R]. Wales:Dept. of Computer Studies University of Glamorgan.

[49] ABDELMOTY A I, WILLIAMS H M. 1994. Approaches to the Representation of Qualitative Spatial Relationships for Geographic Databases [C] // ABDELMOTY A I, WILLIAMS H M. Advanced Geo-graphic Data Modeling:Spatial Data Modeling and Query Language for 2D and 3D Applications. Delft, The Netherlands:204-216.

[50] AGRAWAL R, IMIELINSKI T, SWAMI A. 1993. Mining Association Rules between

Sets of Items in Large Databases: Proc. of 1993 ACM-SIGMOD Int. Conf. Management of Data, Washington DC, May 1993 [C/OL]. [2005-8-5]. http://rakesh.agrawal-family.com/papers/sigmod93assoc.pdf.

[51] AGRAWAL R, SRIKANT R. 1994. Fast Algorithms for Mining Association Rules: Proc. 1994 Int. Conf. VLDB, Santiago, Chile, Sept. 1994[C/OL] [2008-8-8]. http://rakesh.agrawal-family.com/papers/vldb94apriori.pdf.

[52] ALLEN J F. 1983. Maintaining Knowledge about Temporal Intervals [J]. Communications of the ACM, 26(11):832-843.

[53] ARMSTRONG M A. 1997. 基础拓扑学[M]. 孙以丰,译. 北京:北京大学出版社.

[54] BEAUBOEF T, LADNER R, PETRY F. 2004. Rough Set Spatial Data Modeling for Data Mining [J]. International Journal of Intelligent Systems, 19:567-584.

[55] BENNETT B. 1997. Logical Representations for Automated Reasoning about Spatial Relationships [D]. Leeds:School of Computer Studies, University of Leeds.

[56] BENNETT B. 2000. Logics for Topological Reasoning [R]. Birmingham: ESSLLI Summer School, University of Birmingham.

[57] BJORKE J T. 2004. Topological Relations between Fuzzy Regions: Derivation of Verbal Terms [J]. Fuzzy Sets and Systems, 141:449-469.

[58] BRIGGS R. 1973a. On the Relationship between Cognitive and Objective Distance [C] // PREISER W F E. Environmental Design Research. Stroudsburg:Hutchinson and Ross Inc,:186-192.

[59] BRIGGS R. 1973b. Urban Cognitive Distance [C]//DOWNS R M, STEA D. Image and Environment. Chichago:Aldine Publishing Company:361-388.

[60] BRINKHOFF T, KRIEGEL H P, SCHNEIDER R. 1993. Comparison of Approximations of Complex Objects Used for Approximation-based Query Processing in Spatial Database Systems:Ninth International Conference on Data Engineering[C]. Vienna:IEEE Press:81-90.

[61] CADWALLADER M T. 1979. Problems in Cognitive Distance Implications for Cognitive Mapping [J]. Environment and Behavior, 11(4):559-576.

[62] CHANG K-T. 2002. Introduction to Geographic Information Systems[M]. New York: The McGraw-Hill Companies, Inc..

[63] CHANG S K, LI Y. 1988. Representation of Multi-Resolution Symbolic and Binary Pictures Using 2DH Strings[C] // Anon. 1988 IEEE Workshop on Lang for Autom Symbiotic and Intell Rob: 1988 IEEE Workshop on Languages for Automation, Symbiotic and Intelligent Robotics. Piscataway,NJ,United States: IEEE: 190-195.

[64] CHANG S K, SHI Q Y, YAN C W. 1987. Iconic Indexing by 2-D Strings [J]. IEEE Transaction on Pattern Analysis and Machine Intelligence, PAMI-9 (3): 413-427.

[65] CHEN Jun, JIANG Jie. 2000. An Event-based Approach to Spatial-Temporal Data Modeling in Land Subdivision Systems [J]. GeoInformatica, 4(4):387-402.

[66] CHEN Jun, LI Cheng Ming, LI Zhi Lin, et al. 2001. A Voronoi-based 9-Intersection Model forspatial Relations [J]. International Journal of Geophysical Information Sciences, 15(3): 201-220.

[67] CHRISTOPHER B J. 1995. Map Generalization with a Triangulated Data Structure [J]. Cartography and Geographic Information Systems, 22(4):317-331.

[68] CLARKE B L. 1981. A Calculus of Individuals Based on 'Connection' [J]. Notre Dame Journal of Formal Logic, 23(3):204-218.

[69] CLARKE B L. 1985. Individuals and Points [J]. Notre Dame Journal of Formal Logic, 26(1):61-75.

[70] CLEMENTINI E, DI FELICE P, van OOSTEROM P. 1993. A Small Set of Formal Topological Relationships Suitable for End-user Interaction [C] // ABEL D, OOI B C. Proc. 3rd International Symposium on Large Spatial Database, Lecture Notes in Computer Science No. 692. Singapore:Springer-Verlag:277-295.

[71] CLEMENTINI C. 2005. A Model for Uncertain Lines [J]. Journal of Visual Languages and Computing, 16:271-288.

[72] CLEMENTINI E, DI FELICE P. 1995a. A Comparison of Methods for Representing Topological Relationships [J]. Information Sciences, 3:149-178.

[73] CLEMENTINI E, DI FELICE P. 1996. A Model for Representing Topological Relationships between Complex Geometric Features in Spatial Databases [J]. Information Sciences, 90:121-136.

[74] CLEMENTINI E, DI FELICE P. 1997a. Approximate Topological Relations [J]. International Journal of Approximate Reasoning, 16:173-204.

[75] CLEMENTINI E, DI FELICE P. 1998. Topological Invariants for Lines [J]. IEEE Transactions on Knowledge and Data Engineering, 10(1):38-54.

[76] CLEMENTINI E, DI FELICE P. 2001. A Spatial Model for Complex Objects with a Broad Boundary Supporting Queries on Uncertain Data [J]. Data and Knowledge Engineering, 37:285-305.

[77] CLEMENTINI E, DI FELICE P, CALIFANO G. 1995b. Composite Regions in Topological Queries [J]. Information Systems, 20(7):579-594.

[78] CLEMENTINI E, DI FELICE P, HERNÁNDEZ D. 1997b. Qualitative Representation of Positional Information [J]. Artificial Intelligence, 95:317-356.

[79] CLEMENTINI E, DI FELICE P, KOPERSKI K. 2000. Mining Multiple-level Spatial Association Rules for Objects with a Broad Boundary [J]. Data & Knowledge Engineering, 34:251-270.

[80] CLEMENTINI C, SHARMA J, EGENHOFER M J. 1994. Modeling Topological Spatial Relations Strategies for Query Processing [J]. Coput. & Graphics, 18(6):815-822.

[81] COHN A G, BENNET T, GOODAY J, et al. 1997. Qualitative Spatial Representation and Reasoning with the Region Connection Calculus [J]. GeoInformatica, 1:275-316.

[82] COHN A G, RANDELL D A, CUI Z. 1995. Taxonomies of Logically Defined Qualitative Spatial Relations [J]. Internat. J. Human-Comput. Stud. , 43(5-6):831-846.

[83] COUCLELIS H. 1996. Towards an Operational Typology of Geographic Entities with Ill-Defined Boundaries [C]// BURROUGH P A, FRANK A U. Geographic Objects with Indeterminate Boundaries, London:Taylor and Francis:45-56.

[84] CUI Z, COHN A G, RANDELL D A. 1992. Qualitative Simulation Based on a Logical Formalism of Space and Time: Proceedings AAAI-92 [C/OL]. [2006-11-2]. http://eprints.kfupm.edu.sa/60188/1/60188.pdf.

[85] CUI Z, COHN A G, RANDELL D A. 1993. Qualitative and Topological Relationships in Spatial Databases [C] // ABEL D J, OOI B C. Third International Symposium on Advances in Spatial Database, SSD'93, Lecture Notes in Computer Science. Singapore: Springer, 692:296-315.

[86] DOWN R M, STEA D. 1973. Cognitive Maps and Spatial Behavior: Process and Products [C]// DOWNS R M, STEA D. Image and Environment: Cognitive Mapping and Spatial Behavior. Chicago:Aldine, Pub. Co. :8-26.

[87] DUBOIS D, JAULENT M C. 1987. A General Approach to Parameter Evaluation in Fuzzy Digital Pictures [J]. Pattern Recognition Letters, 6:251-259.

[88] DUTTA S. 1989. Qualitative Spatial Reasoning: a Semi-quantitative Approach Using Fuzzy Logic [C]// BUCHMANN A, GUTHER O, SMITH T R, et al. Symposium on the Design and Implementation of Large Spatial Databases, Lecture Notes in Computer Science. New York: Springer-Verlag,409:345-364.

[89] DUTTA S. 1989. Approximate Reasoning with Temporal and Spatial Concepts [R]. Berkeley:Dept. of Computer Science, University of California at Berkeley.

[90] EGENHOFER M J. 1993. A Model for Detailed Binary Topological Relationships [J]. Geomatica, 47(3-4):261-273.

[91] EGENHOFER M J. 1994a. Deriving the Composition of Binary Topological Relations [J]. Journal of Visual Languages and Computing, 5(2):133-149.

[92] EGENHOFER M J. 1994b. Pre-processing Queries with Spatial Constrains [J]. Photogrammetric Engineering & Remote Sensing, 60(6):783-970.

[93] EGENHOFER M J, AI-TAHA K K. 1992. Reasoning about Gradual Change of Changes of Topological Relationships [C] // FRANK A U, CAMPARI I, FORMENTINI U. Theories and Models of Spatio-temporal Reasoning in Geographic Space, Lecture Notes in Computer Science. Berlin: Springer, 639:196-219.

[94] EGENHOFER M J, FRANZOSA R D. 1991a. Point-Set Topological Spatial Relations [J]. Internat. J. Geographical Inform. Systems, 5(2):161-174.

[95] EGENHOFER M J, FRANZOSA R D. 1995. On the Equivalence of Topological Relations [J]. International Journal of Geographic Information Systems, 9(2):133-152.

[96] EGENHOFER M J, HERRING J R. 1991b. Categorizing Binary Topological Relation-

ships between Regions, Lines, and Points in Geographic Databases [R]. Orono, ME: Department of Surveying Engineering, University of Maine.

[97] FRANK A U. 1992. Qualitative Spatial Reasoning about Distances and Directions in Geographic Space [J]. Journal of Visual Languages and Computing, 3:343-371.

[98] FRANK A U. 1996. Qualitative Spatial Reasoning: Cardinal Direction as an Example [J]. International Journal of Geographic Information Systems, 10(3):269-290.

[99] FRANZ A. 1991. Voronoi Diagram: a Survey of a Fundamental Geometric Data Structure [J]. ACM Computing Surveys, 23(3):345-405.

[100] FREKSA C. 1992. Temporal Reasoning Based on Semi-Intervals [J]. Artificial Intelligence, 54:199-227.

[101] FREKSA C, ROHRIG R. 1993. Dimensions of Qualitative Spatial Reasoning [R]. Germany:Dept. of Computer Science, University of Hamburg.

[102] GARLING T, BOOK A, LINDBERG E, et al. 1991. Evidence of Response-Bias Explanation of Non-Euclidean Cognitive Maps [J]. Professional Geographer, 43(2):143-149.

[103] GATRELL A C. 1983. Distance and Space: a Geographical Perspective [M]. Oxford: Clarendon Press.

[104] GOODCHILD M F. 1995. Future Directions for Geographic Information Science [J]. Geographic Information Science, 1(1):1-7.

[105] GOODCHILD M F, GOPAL S. 1990. The Accuracy of Spatial Databases [M]. Basingstoke: Taylor and Francis.

[106] GORE A. 1998. The Digital Earth: Understanding Our Planet in the 21th Century [R]. Los Angeles: California Science Center.

[107] GOYAL R. 2000. Similarity Assessment for Cardinal Directions between Extended Spatial Objects [D]. Orono, ME:the University of Maine.

[108] GOYAL R, EGENHOFER M J. 2001. Similarity of Cardinal Directions [C] // JENSEN C, SCHNEIDER M, SEEGER B, et al. Seventh International Symposium on Spatial and Temporal Database. Los Angeles, CA, Lecture Notes in Computer Science, 2121. Berlin:Springer-Verlag:36-55.

[109] GUESGEN H W. 1989. Spatial Reasoning Based on Allen's Temporal Logic. International Computer Science Institute Technical Report TR-89-049 [R]. Berkeley,CA:International Computer Science Institute.

[110] HAN J, FU Y. 1995. Discovery of Multiple-level Association Rule from Large Databases: Proceedings of the International Conference on Very Large Databases[C]. San Francisco: Morgan Kaufmann Publishers Inc: 20-31.

[111] HERNÁNDEZ D. 1990. Relative Representation of Spatial Knowledge: the 2-D case [C] // MARK D M, FRANK A U. Cognitive and Linguistic Aspects of Geographic Space. Las Navas del Marqués:NATO ASI: 373-385.

[112] HERNÁNDEZ D. 1994. Qualitative Representation of Spatial Knowledge: Lecture

Notes in Artificial Intelligence 804[C]. Berlin：Springer-Verlag.

[113] HERNÁNDEZ D, CLEMENTINI E, DI FELICE P. 1995. Qualitative Distances：Proceedings of COSIT'95, LNCS[C]. Springer-Verlag, Berlin.

[114] HONG J H. 1994. Qualitative Distance and Direction Reasoning in Geographic Space [D]. Orono, ME：University of Maine.

[115] HONG J H, EGENHOFER M J, FRANK A U. 1995. On the Robustness of Qualitative Distance and Direction-Reasoning：Proceedings of AutoCarto 12 [C/OL]. [2008-10-9]. ftp：// ftp. geoinfo. tuwien. ac. at/frank/2013. Robustness_af. pdf.

[116] JUNGERT E. 1988. Extended Symbolic Projection Used in a Knowledge Structure for Spatial Reasoning：Proceedings of 4th BPRA Conference on Pattern Recognition [C/OL] [2007-1-2]. http：// www. springerlink. com/content/y52q4723k37318p6/fulltext. pdf.

[117] JUNGERT E. 1992. The Observer's Point of View：an Extension of Symbolic Projections [C]// FRANK A U, CAMPARI I, FORMENTINI U. Theories and Methods of Spatio-temporal Reasoning in Geographic Space：Lecture Notes in Computer Science 639. Berlin：Springer-Verlag：179-195.

[118] JUNGERT E, CHANG S K. 1989. An Algebra for Sympolic Image Manipulation and Transformation [C]// KUNII T L. Visual Database Systems：Proceedings of the IFIP TC 2/WG 2. 6 Working Conference on Visual Database Systems. Tokyo, Japan：Elserier Science Publishers：301-317.

[119] KIRASIC K C, ALLEN G L, SIEGEL A W. 1984. Expression of Configurational Knowledge of Large-scale Environment [J]. Environment and Behavior, 16：687-712.

[120] KOPERSKI K, HAN J. 1995. Discovery of Spatial Association Rules in Geographic Information Databases：Proceedings of the Fourth International Symposium on Large Spatial Databases[C]. [2007-2-1]. http：// www. math. unipd. it/~dulli/corso04/ssd95. pdf.

[121] KOSHIZUKA T, KURITA O. 1991. Approximate Formulas of Average Distance Associated with Regions and their Applications to Location Problems [J]. Mathematical Programming：Session B, 52：99-123.

[122] KUIPERS B. 1990a. Commonsense Knowledge of Space：Learning from Experience[C]// CHEN S S. Advances in Spatial Reasoning. Norwood, New Jersey, USA：Ablex Publishing Corporation：199-206.

[123] KUIPERS B. 1990b. Modeling Spatial Knowledge [C]// CHEN S S. Advances in Spatial Reasoning. Norwood, New Jersey, USA：Ablex Publishing Corporation：171-198.

[124] KUIPERS B, LEVITT T. 1990. Navigation and Mapping in Large-scale Space [C]// CHEN S S. Advances in Spatial Reasoning. Norwood, New Jersey, USA：Ablex Publishing Corporation：207-251.

[125] LADNER R, PETRY F E, COBB M A. 2003. Fuzzy Set Approaches to Spatial Data

Mining of Association Rules [J]. Transactions in GIS, 7(1):123-138.

[126] LIU Xin, LIU Wen Bao, LI Cheng Ming. 2008a. Qualitative Description and Reasoning of Topological Relation in Three-Dimensional GIS [C]. Anon. 3rd International Conference on Innovative Computing Information and Control, ICICIC'08. Piscataway, NJ 08855-1331, United States: Inst. of Elec. and Elec. Eng. Computer Society.

[127] LIU Xin, LIU Wen Bao, LI Cheng Ming. 2008b. Qualitative Representation and Reasoning of Combined Direction and Distance Relationships in Three-dimensional GIS [C]. Anon. International Conference on Computer Science and Software Engineering, CSSE 2008. Piscataway, NJ 08855-1331, United States: Inst. of Elec. and Elec. Eng. Computer Society.

[128] LUNBERG U. 1973. Emotional and Geographical Phenomenon in Psychophysical Research[C] // DOWNS R M, STEA D. Image and Environment. Chichago: Aldine Publishing Company: 322-337.

[129] MAKI R H. 1981. Categorization and Distance Effect with Spatial Linear Orders [J]. Journal of Experimental Psychology, 7(1):15-32.

[130] MCDERMOTT D, DAVIS E. 1984. Planning Routes Through Uncertain Territory [J]. Artificial Intelligence, 22:107-156.

[131] MCKAY R I, PEARSON R A, WHIGHAM P A. 1997. Learning Spatial Relationships: Some Approaches: Proceedings of Geo Computation'97 and SIRC'97 [C/OL]. [2006-5-15]. http: // citeseerx. ist. psu. edu/viewdoc/download? doi = 10. 1. 1. 106. 7428&rep=rep1&type=pdf.

[132] MCNAMARA D V, DAVIS E. 1984. Planning Routes through Uncertain Territory [J]. Artificial Intelligence, 22:107-156.

[133] MCNAMARA T P, LESUEUR L L. 1989. Mental Representations of Spatial and Nonspatial Relations [J]. The Quarterly Journal of Experimental Psychology, 41A:215-233.

[134] MENNIS J, LIU J W. 2005. Mining Association Rules in Spatio-Temporal Data: an Analysis of Urban Socioeconomic and Land Cover Change [J]. Transactions in GIS, 9(1):5-17.

[135] MOLEANAAR M. 1992. A Topology for 3D Vector Maps[J]. ITC Journal, (1):25-33.

[136] MOLEANAAR M. 1989. Towards a Geographic Information Theory [J]. ITC Journal, (1):5-11.

[137] MONTANARI A, PERNICI B. 1993. Temporal Reasoning[C] // TANSEL A U, CLIFFORD J, GADIAETAL S. Temporal Databases: Theory, Design, and Implementation. Redwood, CA: The Benjaming/Cummings Publishing Company, Inc. : 534-562.

[138] MONTELLO D R. 1991. The Measurement of Cognitive Distance: Methods and Construct Validity [J]. Journal of Environment Psychology, 11(2):101-122.

[139] PAPADIAS D, SELLIS T. 1992. Spatial Reasoning Using Symbolic Arrays [C] // FRANK A U, CAMPARI I, FORMENTINI U. Theories and Methods of Spatio-temporal Reasoning in Geographic Space: International Conference GIS-From Space to Territory, Pisa, Italy, Lecture Notes in Computer Science 639. Berlin: Springer-Verlag: 153-161.

[140] PAPADIAS D, SELLIS T. 1994. Qualitative Representation of Spatial Knowledge in Two-dimensional Space [J]. The VLDB Journal, 3(4):479-516.

[141] PAPADIAS D, THEODORIDIS Y, SELLIS T, et al. 1995. Topological Relations in the World of Minimum Bounding Rectangles: a Study with R-Trees:Proceedings of the ACM SIGMOD International Conference on Management of Data[C]. San Jose, Calif. : ACM Press:92-103.

[142] PAPADIAS D, KARACAPILIDIS N, ARKOUMANIS D. 1999. Processing Fuzzy Spatial Queries: a Conguration Similarity Approach [J]. International Journal of Geographical Information Science, 13(2):93-118.

[143] PEUQUET D J. 1992. An Algorithm for Calculating Minimum Euclidean Distance between Two Geographical Features [J]. Computers and Geosciences, 18(8):989-1001.

[144] PEUQUET D J, ZHAN C X. 1987. An Algorithm to Determine the Directional Relationship between Arbitrarily-shaped Polygons in the Plane [J]. Pattern Recognition, 20(1):65-74.

[145] RANDELL D A, DAVID A R, CUI Z, COHN A G. 1992. A Spatial Logic Based on Regions and Connection: Proc. 3rd Int. Conf. on Principles of Knowledge Representation and Reasoning [C/OL]. [2006-10-7] http://citeseerx.ist.psu.edu/viewdoc/download?doi=10.1.1.35.7809&rep=rep1&type=pdf.

[146] RENZ J. 1998. A Canonical Model of the Region Connection Calculus:Proc. of the 6th International Conference on Principles of Knowledge Representation and Reasoning[C/OL]. [2005-11-3]. http://users.rsise.anu.edu.au/~jrenz/papers/renz-jancl02.pdf.

[147] SADALLA E K, STAPLIN L J. 1980a. The Perception of Traversed Distance Intersections [J]. Environment and Behavior, 12(2):167-182.

[148] SADALLA E K, STAPLIN L J. 1980b. An Information Storage Model for Distance Cognition [J]. Environment and Behavior, 12(2):183-193.

[149] SANTOS M Y, AMARAL L A. 2005. Geo-spatial Data Mining in the Analysis of a Demographic Database [J]. Soft Computing, 9:374-384.

[150] SCHLIEDER C. 1995. Reasoning about Ordering:A Theoretical Basis for GIS International Conference COSIT'95 [C/OL]. [2006-10-9] http://www.springerlink.com/content/fgg4g71004655854/fulltext.pdf.

[151] SERGENT J. 1991. Judgments of Relative Position and Distance on Representation of Spatial Relations [J]. Journal of Experimental Psychology, 17(3):762-780.

[152] SHARMA J. 1996. Integrated Spatial Reasoning in Geographic Information Systems:

Combining Topology and Direction [D]. Orono, ME: University of Maine.

[153] SHARMA J, FLEWELLING D M. 1995. Inferences from Combined Knowledge about Topology and Directions [C] // HERRING J R, EGENHOFER M J. 4th International Symposium on Large Spatial Databases, SSD'95, Lecture Notes in Computer Science. Berlin: Springer-Verlag.

[154] SIMMONS R. 1990. Commonsense Arithmetic Reasoning [C] // WELD D S, KLEER J D. Readings in Qualitative Reasoning about Physical Systems. San Mateo, California: Morgen Kaufmann Publishers, Inc.: 337-343.

[155] STRUSS P. 1990. Problems of Interval-Based Qualitative Reasoning [C] // WE^{LD} DS, KLEER JD. Readings in Qualitative Reasoning about Physical Systems. San Mateo, California: Morgan Kaufmann Publishers, Inc.: 288-305.

[156] THORNDYKE P W. 1981. Distance Estimation from Cognitive Maps [J]. Cognitive Psychology, 13:526-550.

[157] TIMPF S. 1992. A Conceptual Model of Way Finding Using Multiple Levels of Abstraction: Proc. of the GIS-form Space to Theory: Theories and Methods of Spatio-temporal Reasoning[C/OL]. [2007-2-4]. http://www.spatial.maine.edu/~max/wayFinding-Abstraction.pdf.

[158] VILAIN M, KAUTZ H, VAN BEEK P. 1990. Constraint Propogation Algorithms for Temporal Reasoning: a Revised Report [C] // WELD D S, KLEER J D. Readings in Qualitative Reasoning about Physical Systems. San Mateo, California: Morgan Kaufmann Publishers, Inc.:373-381.

[159] WANG F J. 2003. Handling Grammatical Errors, Ambiguity and Impreciseness in GIS Natural Language Queries [J]. Transactions in GIS, 7(1):103-121.

[160] WOLTER F, ZAKHARYASCHEV M. 2000. Spatial Reasoning in RCC-8 with Boolen Region Terms: Proc. Of the 4th European Conference on Artificial Intelligence, ECAI 2000[C/OL]. [2006-7-2]. http://citeseerx.ist.psu.edu/viewdoc/download?doi=10.1.1.38.8880&rep=rep1&type=pdf.

[161] WRIGHT D J, GOODCHILD M F. 1997. Data from the Deep: Implications for the GIS Community [J]. International Journal of Geographical Information Systems, 11(6):523-528.

[162] ZHAN F B. 1998. Approximate Analysis of Binary Topological Relations between Geographic Regions with Indeterminate Boundaries [J]. Soft Computing, 2:28-34.

[163] ZLATANOVA S. 2000. On 3D Topological Relationships: Proc. of the 11th International Workshop on Database and Expert System Applications. Greenwich, London[C/OL]. [2008-2-1]. http://www.gdmc.nl/zlatanova/thesis/html/refer/ps/sz_asdm.pdf.